VHDL Starter's Guide

Sudhakar Yalamanchili

School of Electrical and Computer Engineering
Georgia Institute of Technology

PRENTICE HALL
Upper Saddle River, New Jersey 07458

Library of Congress Cataloging-in-Publication Data

Yalamanchili, Sudhakar
    VHDL starter's guide / Sudhakar Yalamanchili
    p. cm.
    Includes bibliographical references and index.
    ISBN: 0-13-519802-X (pbk.)
        1. VHDL (Hardware description language) 2. Very high speed
integrated circuits—Design and construction—Data processing.
I. Title.
TK7885.7.Y35 1998   97–18139
821.39'5–DC21   CIP

Acquisitions editor: *TOM ROBBINS*
Editor-in-chief: *MARCIA HORTON*
Managing editor: *BAYANI MENDOZA DE LEON*
Director of production and manufacturing: *DAVID W. RICCARDI*
Production editor: *KATHARITA LAMOZA*
Cover design: *BRUCE KENSELAAR*
Manufacturing buyer: *DONNA SULLIVAN*
Editorial assistant: *NANCY GARCIA*

© 1998 by Prentice-Hall, Inc.
Upper Saddle River, New Jersey 07458

Printed in the United States of America

10 9 8

ISBN   0-13-519802-X

Prentice-Hall International (UK) Limited, *London*
Prentice-Hall of Australia Pty. Limited, *Sydney*
Prentice-Hall Canada Inc., *Toronto*
Prentice-Hall Hispanoamericana, S.A., *Mexico*
Prentice-Hall of India Private Limited, *New Delhi*
Prentice-Hall of Japan, Inc., *Tokyo*
Prentice-Hall Asia Pte. Ltd., *Singapore*
Editora Prentice-Hall do Brasil, Ltda., *Rio de Janerio*

To

Padma, Suchitra, and Sunil

Contents

Preface

If I hear, I forget
If I see, I remember
If I do, I understand
                    —Proverb

Over the years, hardware description languages have evolved to aid in the description, modeling, and design of digital systems. However, it has not been until recently that their use has accelerated. The confluence of advances in design automation, exponentially increasing computer power at the desktop, and the emergence of an IEEE Standard for VHDL has contributed to the growth. The rapid development and use of the IEEE Standard VHDL language in industry makes it imperative that we provide opportunities for Electrical and Computer Engineering students to learn the language and become proficient in its application.

There are many excellent texts on the description of the VHDL language and its use for the purpose of building accurate models of complex digital systems. These texts have been largely complete treatments of the language and generally have been written for practicing engineers and more recently also for use in graduate courses. This text has a different goal. It is not intended to be a comprehensive VHDL language reference. Rather, it is intended to provide an introduction to the *basic* language concepts, and a framework for *thinking about* the structure and operation of VHDL programs. Programming idioms from conventional programming languages by themselves are insufficient for productively learning to apply hardware description languages such as VHDL. Through simulation and laboratory exercises, students can very quickly come up to speed in building useful, non-trivial models of digital systems. As their experience grows, so will their need for more comprehensive information and modeling techniques, which can be found in a variety of existing texts and the standard language reference manual.

The VHDL language can be used with several goals in mind. It may be used for the synthesis of digital circuits, verification and validation of digital designs, test vector generation for testing circuits, or simulation of digital systems. The use of the language for each of these purposes will emphasize certain of its aspects. This text is intended to support a first look at the language and as such I have decided to adopt the application to the simulation of digital systems in introducing the basic language concepts. Digital logic synthesis is an alternative, widely used application of the language. A good introduction to the use of VHDL for logic synthesis can be found in several texts (see [10,12]).

Intended Audience

The style of the book is motivated by the need for a companion text for sophomore- and junior-level texts used in courses in digital logic and computer architecture. The ability to construct simulation models of the building blocks studied in these courses is an invaluable teaching aid. VHDL is a complex language that could easily be worthy of a course in its own right. However, curricula are usually strapped for credit hours, and devoting a course to teach VHDL would mean eliminating existing material. The style of this text is intended to permit integration of the basic concepts underlying VHDL into existing courses without necessitating additional credit hours or courses for instruction.

Students learn the most about digital systems if they have to build them, in this case using simulation models. I have found it valuable in my own courses to provide concurrent laboratory exercises that reinforce foundational material taught in the classroom. The scope and complexity of hardware laboratory exercises are limited by the available time in a semester or quarter-long course. The VHDL language provides an opportunity for students to experiment with larger designs than would be feasible in a hardware laboratory, using a development environment that is rapidly being adopted throughout the industrial community. In the case of VHDL I would like an approach that would enable students to adjust quickly to the basic language concepts such that they could construct models of basic logic and computer architecture components productively in sophomore- and junior-level courses. The full power of the language is not necessary at this point, nor should it be. As they progress to more advanced courses and their needs grow, students would be able to use productively the more advanced language features and their corresponding texts as references.

This book attempts to develop an intuition and a structured way of thinking about VHDL models without necessarily spending a great deal of time on advanced language features. Students should be able to learn enough quickly through exercises and association with classroom concepts to be able to construct useful models. This book strives to fill this need. During development, portions of this book have been utilized in a two-course sequence on computer architecture at Georgia Tech. Sophomore students start with no background in VHDL but with a good background in high-level programming languages such as C or Pascal. By the end of the second quarter they will have built a model of a pipelined RISC processor with hazard detection, data forwarding, and branch prediction. Early in the second quarter VHDL makes the transition from a new language to a tool for studying computer architecture. The goal is to integrate VHDL into the curriculum early in a manner that strengthens the learning of the concepts while concurrently providing training in the use of VHDL simulation tools.

Style of the Book

In order to fill the need for a companion text for computer architecture courses and an early introduction to the basic language concepts the book must satisfy several criteria. First, it must relate VHDL concepts to those already familiar to the student. Students learn best when they can relate new concepts to ones with which they are already familiar. In this case we rely on concepts from the operation of digital circuits. Language features are motivated by the need to describe specific aspects of the operation of digital circuits, for example, events, propagation delays, and concurrency. Second, each language feature must be accompanied by examples. Simulation exercises address one or more VHDL modeling concepts. To keep with the idea of a companion book tutorials for two VHDL simulators are provided in the Appendices. Finally, the text must be *prescriptive*. Chapter 3, Chapter 4, and Chapter 5 each provide a prescription for writing classes of VHDL models. This is not intended to produce the most efficient models, but serves the purpose of rapidly bringing the student to a point where he or she can construct useful simulation models for instructional purposes. By enabling a look at the detailed operation of digital systems, VHDL reinforces the foundational concepts taught in the classroom. At this point, students begin to think about alternative, more creative, and often more efficient, approaches to constructing the models.

The approach taken is a bit unusual in that I do not begin with a discussion and presentation of language syntax and constructs, that is, identifiers, operators, and so forth. In fact, the book presents core constructs via examples and the syntax is not presented until late in the text with the notion that this chapter will be used more as a reference. The premise is that readers who have had experience with programming in high-level languages simply need an accurate syntactic reference to these language constructs. The road to building useful models is built on an understanding of how we can describe those aspects of hardware systems that require constructs that are not typically found in traditional programming language definitions, such as signals and the concept of time. This is where the bulk of this book is focused. As a result, a syntactic reference to the core programming language features has been reduced to a single chapter. The goal is to focus on the concepts underlying the VHDL language and its use in simulation. If we can capture the novel features of the language in a manner that appeals to the reader's intuition and is based on thinking in hardware or systems terms, then a reference to the syntax of core constructs is sufficient to get students started in building useful models. My hope is that this text can apply this approach successfully and fulfill the goal of getting students at the sophomore level excited about the use of such languages in general and the evolving design methodologies in particular. They are then capable of more rapidly expanding their understanding to the full scope of the language.

Organization of the Text

As a result of this view, the text starts with concepts from the operation of digital circuits. This text assumes that the reader is comfortable with introductory digital logic and programming in a block-structured, high-level language such as Pascal or C. The subsequent chapters introduce various VHDL constructs as representations of the operational and physical attributes of digital systems. An association is made between the physical phenomena found in digital circuits and their representation within the VHDL language. Though VHDL is often criticized as a bulky and complex language in its entirety, these associations are intuitive and therefore easy to pick up. Once the basic concepts are understood a good syntactic reference to the language is sufficient for the students to be able to build models quickly of digital circuits including higher-level objects such as register files, ALUs, and simple datapaths.

Chapter 3 is perhaps the most important chapter in the book. The basic attributes of digital systems are associated with language features and an overall structure of a VHDL model is developed. Chapter 4 and Chapter 5 build on this basic structure to introduce the concepts of processes, hierarchy, and abstraction. Each of these chapters provide simulation exercises that can be exercised by the student to reinforce the concepts.

Completion of the simulation exercises provides the student with an ability to proceed to more complex language features and productive use of any of the existing comprehensive VHDL language texts. In our curriculum, the students proceed to senior design projects where they describe systems in VHDL, synthesize their designs, and emulate the resulting designs using hardware emulators. Chapters 6 through 8 present language features as additional functionality that serve a specific purpose, for example, input–output and procedures. By this time students are able to write and simulate simple models. Thus, students can concentrate on each of these new features and their utilities. A syntactic reference to the common language features is provided in Chapter 9.

Several appendices have been added to support the material in the text. For example, tutorials for two commonly used simulators are provided. The tutorials include brief descriptions of common simulator operations such as examination of the event queue, tracing signals, and generating input stimuli. Another appendix provides a detailed template for a VHDL model illustrating the relative ordering of program constructs. This can serve as a handy reference toward the end of the student's experience. A third appendix provides a description of a model of the SPIM processor datapath described in "Computer Organization & Design: The Hardware Software Interface" by D. Patterson and J. Hennessey [9]. Students in a first quarter architecture course can modify this datapath by adding instructions or modifying the controller to implement a more complex state machine. The model can be used in a second quarter computer architecture course as the starting point for constructing a pipelined datapath with hazard detection, forwarding, and simple branch prediction.

By providing a companion text for gradually enabling students to learn VHDL "as they go," I hope that the text will enable more curricula to productively introduce VHDL early in the curriculum while serving as a teaching aid for classroom material. Students are thus prepared for advanced courses dealing with state-of-the-art techniques such as rapid prototyping of digital systems, high-level synthesis, and advanced modeling.

Caveats

This book is based on the 1987 VHDL standard. The language has since been revised pro- **VHDL'93** ducing the 1993 VHDL standard. Those features that are only available in VHDL'93 are explicitly identified in the text and marked in the margin as shown here. However, we do not cover the full set of VHDL'93 revisions. This text is based on the idea that at the sophomore and junior level it is best to limit coverage to those features of the language that can be used to build useful simulation models which provide insights into the behavior of digital systems, particularly at the architectural level. Students can subsequently can move onto a broader coverage of the language in many excellent textbooks, some of which are listed in the bibliography. As a result, the reader should be aware that complete coverage of the language features is not provided in this text. Furthermore, early in the text the IEEE 1164 standard data types are introduced and utilized in all of the examples rather than staying with the relatively simpler **bit** and **bit_vector** types. We hope that this will expose students to standard practice and not compromise the goals of focusing on core language features and types.

Finally, all of the models shown in the text have been compiled and tested under Workview Office version 7.10 from Viewlogic Systems. In some examples the package `std_logic_arith.vhd` was used. While this package is generally available in the library `IEEE`, readers should be aware that in a different installation this package could be in another directory. In our examples, I utilized the package that was compiled into the library `WORK`.

Acknowledgments

This book grew out of a perceived need in the classroom, the encouragement of colleagues, the participation and feedback from students, and the never-ending accommodations of my family. It is a pleasure to acknowledge the contributions of the many people who contributed to its generation.

I am particularly grateful to John Uyemura, who provided the initial impetus for the book and his continued support and input throughout the project. His experience with how students learn was invaluable in shaping the approach employed in the text. I am also grateful to Tom Robbins of Prentice Hall for his patience with the ever-changing deadlines, his commitment to education, and many insightful conversations with him regarding the teaching and learning process.

I am especially grateful to those colleagues who were generous with their time in reading early drafts of the manuscript and forthcoming with their comments and suggestions. Todd Carpenter, Vijay Madisetti, and Linda Wills took valuable time to comment on early drafts. Jay Schlag, Linda Wills and Abhijit Chatterjee utilized early drafts in a first course in Computer Architecture taught at Georgia Tech. I am particularly grateful to numerous sophomore and junior students in CmpE 2510 and CmpE 3510 for their frank and forthright appraisal of the material and the opportunity to gain their feedback. The conscientiousness of the reviewers was evident in the depth of their critique, and improvements over early drafts were largely due to their diligent appraisals in identifying inconsistencies, suggesting illustrative exercises, and improving the presentation style. I am especially grateful to Katharita Lamoza for her tireless editorial efforts in improving the quality of the manuscript and for an education in the presentation of textbook material.

This being my first book, I terribly underestimated the amount of effort in translating teaching in the classroom to communication of concepts and exercises via text. The result was many lost hours, missed engagements, and cancelled plans with my family. This text is dedicated to them as a small measure of my appreciation.

My colleagues and family share any successes from the completion of this text, while any omissions and errors remain solely with the author.

Sudhakar Yalamanchili
Atlanta
July 1997

CHAPTER 1

Introduction

1.1 What is VHDL?

The acronym VHDL stands for the **V**HSIC **H**ardware **D**escription **L**anguage. The acronym VHSIC, in turn, refers to the **V**ery **H**igh **S**peed **I**ntegrated **C**ircuit program. This program was sponsored by the Department of Defense (DoD) with the goals of developing high-speed integrated circuits and involved several major DoD contractors. During the course of this program the need for a standardized representation of digital systems became apparent. A team of DoD contractors was awarded the contract to develop the language and the first version was released in 1985. The language was subsequently transferred to the IEEE for standardization after which representatives from industry, government, and academe were involved in its further development. Subsequently the language was ratified in 1987 and became the IEEE 1076–1987 standard. The language was reballoted after five years and with the addition of new features forms the 1076–1993 version of the language. This text is based on the 1076–1987 standard.

Ever since VHDL became an IEEE standard it has enjoyed steadily increasing adoption throughout the electronic systems CAD community. The DoD requires that VHDL descriptions be delivered for all application-specific integrated circuits (ASICs). In the last few years interoperability between models developed using VHDL environments from different CAD vendors has been improved by the establishment of the IEEE 1164 standard package and other standardization efforts. Practically every major CAD vendor supports VHDL. This status of VHDL as an industry standard provides a number of practical benefits.

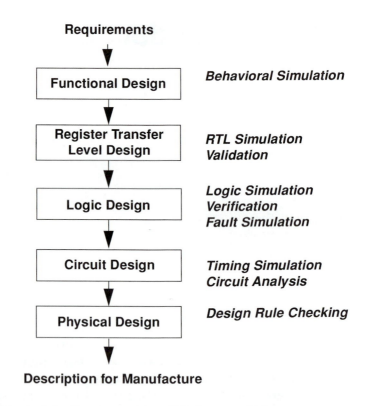

Requirements

Functional Design   *Behavioral Simulation*

**Register Transfer
Level Design**   *RTL Simulation
                               Validation*

Logic Design   *Logic Simulation
                               Verification
                               Fault Simulation*

Circuit Design   *Timing Simulation
                               Circuit Analysis*

Physical Design   *Design Rule Checking*

Description for Manufacture

FIGURE 1-1 Activity flow in top-down digital system design

Conventional procedural programming languages such as C or Pascal typically describe procedures for computing a mathematical function or manipulating data, for example, matrix multiplication or sorting. The execution of the program results in the computation of data values. Similarly, VHDL is a language for describing digital systems. Execution of a VHDL program results in a simulation of the digital system. Although VHDL has been investigated for its use in describing and simulating analog systems, the language is predominantly used in the design of digital systems.

1.2 Digital System Design

The design of digital systems is a process that starts from the specification of requirements and proceeds to produce a functional design that is eventually refined to a physical implementation. Consider the design of an ASIC for processing digital images. An example of the sequence of activities that typically takes place during design is shown in Figure 1-1.

The first step is the specification of the requirements. Such a specification will typically include the performance requirements derived from the number of images to be processed/sec and the operations to be performed on them, as well as interface requirements, cost constraints, and other physical requirements such as size and power dissipation. From these functional requirements, a preliminary high-level functional design can be generated. Simulation is often used at this level to converge to a functional design that can meet the performance requirements. This initial functional design is now refined to produce a more detailed design description at the level of registers, memories, arithmetic units, and state machines. This is the register transfer level (RTL) of the design. Subsequent refinement of this RTL description produces a logic design that implements each of the RTL components. Both RTL and logic level simulation may be used to ensure that the design meets the original specification. Fault simulation can model the effects of expected manufacturing defects as well as faults that may be induced due to the environment. For example, if this image-processing chip is to be flown in a satellite, radiation effects in space can cause devices to change state and lead to single bit errors. If the error rate in the intended orbit is relatively high the design can be modified to accommodate such bit errors using techniques tuned to the particular physical phenomena. Finally, the logic level implementation is transformed into a circuit level implementation and thence to a physical chip layout from which accurate physical properties of the design, such as chip area and power dissipation, can be evaluated. Design rule checks, circuit parameter extraction, and circuit simulation activities can be performed at this level. At each level of this design hierarchy there are components that are used to describe the design. At the higher or more abstract levels we have a smaller number of more powerful components such as adders and memories. At the lower and less abstract levels we have simpler, less powerful components such as gates and transistors. Each level of the design hierarchy corresponds to a *level of abstraction* and has an associated set of activities and design tools that support the activities at this level. Some of the activities at each level are shown in Figure 1-1.

If design errors are discovered at these finer levels of detail changes in the design may be expensive to make, particularly if we have to move back several levels in the design process to correct these errors. This can lead to longer development times and consequently increased cost not to the mention loss of revenues by being late to market. Much of the motivation for the development of hardware description languages in general stems from the evolving economics of the marketplace for electronic systems and the methodologies used to design these systems. With the ability to simulate designs at multiple levels of abstraction, errors can be discovered and corrected early. Hardware description languages such as VHDL provide such a capability.

1.3 The Marketplace

The use of computer-aided design environments in general and hardware description languages in particular is driven by the technology marketplace. After all, these tools and languages are intended to enable the cost-effective and profitable development of electronics products. The key question with respect to the state of the art is: where are the costs, both in terms of costs incurred and revenues lost due to inefficient design paradigms?

The costs incurred are directly a function of the available technology. Both the semiconductor industry and software industry have been in overdrive during the last three decades and show no signs of abatement in the next decade. Memory densities have been quadrupling every 3 years and processor speeds doubling every 18 months. As chip densities increase, new products are becoming available at a faster rate and development cycles are becoming shorter. The importance of time to market is captured in an illustrative manner by Madisetti as shown in Figure 1-2 [7]. For every product there is an optimal time for introduction into the marketplace in a manner that will maximize the revenue generated. Early in the life cycle of the product, market share should rise, peak, and then begin to decline as newer products emerge. If the product is delayed to market, then the trajectory of its performance relative to the maximum may be depicted as shown in the figure. It is very difficult to make up for lost ground. Many organizations measure lost opportunity cost in terms of thousands to tens of thousands of dollars per day in lost revenue. With the

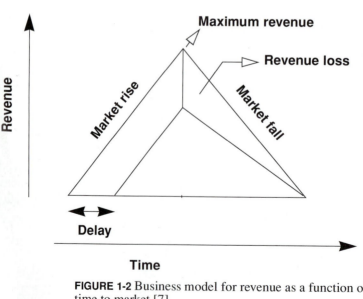

FIGURE 1-2 Business model for revenue as a function of time to market [7]

increasingly faster rate of technology evolution approaches to reduce time to market are critical. New CAD tool environments and hardware description languages are an integral part of any solution.

Given this emphasis on the time-to-market and the resulting push toward shorter design cycles, what are the impediments to doing so? Historically, the CAD tool industry has been driven in a bottom up fashion and has been driven by the development of point tools and processes: design rule checking, layout, verification, and so forth. At the level of chip and board designs, for quite some time there has been an understanding of the specific design problems, which include layout, design rule checking, and test vector generation. Advances have focused on better tools to address these problems spurred by the new challenges of faster, denser, more complex technologies. In contrast, very few tools exist at the architectural level for addressing system design issues. The problems are not as well defined, but the impact can be overwhelming. It has been observed that the first 10%–20% of the design cycle can determine 70%–80% of the final system cost. Further, it has been reported that only 5%–10% of the design cycle time is spent in studying and formulating requirements, while 70% of the manufacturing costs are affected by customer requirements [7].

New methodologies are needed to address the issues of requirements capture, system specification, and early analysis via rapid prototyping. Such new design methodologies are emerging, and VHDL is becoming an integral part of such design approaches.

1.4 The Role of Hardware Description Languages

Traditional design methodologies have been structured around a hierarchy of representations of the system being designed. Distinct representations at differing levels of detail are necessary for the various tasks encountered during design. One of the best-known representations of the different views and levels of abstraction in a digital system is the Y–chart shown in Figure 1-3 [3,13,14] and illustrated through the following example.

Imagine a company DSP, Inc., in the business of designing a next-generation digital signal-processing chip, code name Cyclone. Given the current costs of fabrication and the time window within which the chip must be brought to market to be competitive, we wish to verify prior to fabrication that the chip can support the intended applications. From the design flow shown in Figure 1-1, we know that this behavior can be described at multiple levels of abstraction, that is, at the functional level, RTL level, logic level, and so on. Early in the design behavioral descriptions are necessary so that simulations can ensure that the chip is functionally correct. This functionality may be verified at more than one level of abstraction. The advantage of this approach is that we can make this assessment independent of the many possible physical implementations. Once we have verified the functionality the design can be translated to a structural description comprised of the major components of this chip: memories, registers, arithmetic units, logic units, and so forth. Simulation can be employed again to ensure that the Cyclone structural design correctly

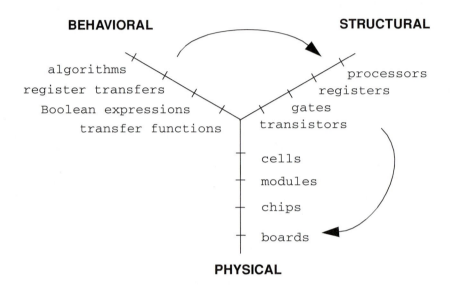

FIGURE 1-3 Design views and corresponding levels of abstraction

performs the intended functions using the components that we have selected. As shown in Figure 1-3, the structural description may also be provided with varying levels of detail. This design can be refined until we translate this description to a physical description that can be further refined to produce a manufacturable specification.

Historically, hardware description languages have been targeted to a certain level of abstraction such as the gate level or register transfer level. Tools have been developed and targeted for tasks at a specific level of the design. For example, some tools may be optimized for implementing state machines that describe data movement at the register transfer level. Other tools may be developed for verification at the gate level by generating test vectors used for final chip verification. The tools at the different levels of abstraction may employ different descriptions of the Cyclone. Such *point tools* are focused on a single aspect of the design and on a single level of abstraction. Often these tools came with their own languages and associated compilers and simulators. However, as chips become more complex and design processes start using an increasingly diverse set of tools, the time taken to move information between tools has become a concern. Recently, estimates of the lost productivity due to incompatibility between design tools has been estimated as high as $4.5 billion dollars/year [7].

Interoperability => The VHDL language provides a set of constructs that can be applied at multiple levels of abstraction and multiple views of the system. This significantly expands the scope of the application of the language and promotes a standardized, portable model of electronic systems. Thus, the complexity of the movement of data and design information between tools is significantly reduced. The net effect is a reduction in the time to design the Cyclone and bring it to market. We may also expect that the reduction in the number of distinct types of tools and languages realizes a reduction in the cost of the design infrastructure and therefore a reduction in the product unit cost.

Technology is a rapidly moving target. We cannot anticipate innovations in technology design styles or products. Therefore an attractive philosophy is to develop a design environment that is independent of technology. The industry is characterized by a number of distinct technologies and associated design styles that target specific points in the continuum of time to market, cost, and performance. For example, the use of programmable logic devices (PLDs) and field programmable gate arrays (FPGAs) has the attributes of low cost and quick time to market. ASIC products incur higher nonrecurring engineering development costs and therefore usually higher product costs. However, they can deliver substantially higher performance. These distinct design styles leads to environments with a distinct set of CAD tools and methodologies.

Technology Independence => The VHDL descriptions of the design are not tied to a specific methodology or target technology. The language is rich enough that it can be used to describe a chip at the instruction set level, register transfer level, or switching transistor level. The major CAD vendors have tools that can synthesize designs to FPGA and complex PLD (CPLD) devices. Design tools for custom and ASIC chips utilize VHDL in their suite to simulate and validate their designs prior to detailed physical design. If the application environment is not very demanding we may choose to synthesize the Cyclone description to a FPGA based implementation. Alternatively we may wish to pursue a higher performance ASIC implementation. Prior to physical design we would wish to have detailed timing information to ensure that the performance requirements of the application can be met. However, the application software developers may simply want a functionally accurate instruction set simulation of the Cyclone chip so that development of application software may begin. Eventually when a detailed physical design is available, the software may be tested on an associated simulation to determine the performance that can be achieved for the applications of interest—two widely differing levels of detail, both easily supported within the same language.

Now let us say that your company develops a model of the Cyclone using CAD tools from vendor Tools'R'Us. Now your company wishes to have a subcontractor use this chip in the design of a second product, say a voice recognition board for personal communicators. Their design environment is completely different, using archaic design tools from StoneAge CAD Tools, Inc. You now need to send them detailed schematics and operational specifications and educate them on the Cyclone design so that they may use this description to design and validate the board level product using their design tools. However, they do support the VHDL language. Since VHDL is a standard, rather than having them reconstruct the design in their environment, the preceding instruction set models of the Cyclone can be directly used and simulated, and will produce identical results using the StoneAge's simulators.

Design Reuse=> We can see libraries of VHDL models of components emerging and being shared across platforms, toolsets, organizations, and technical groups. Engineers working on a large design can be independently designing subsystems with considerably less concern for design environment or design tool compatibly issues.

Part of the difficulty with a board design is that the software for processing the data streams cannot be tested until hardware is available. But what if we have detailed hardware descriptions of the components on the board that behave exactly as the Cyclone chip to the level of detail of a clock cycle? We could simulate the system to a sufficient level of detail so that the simulator could take the place of the hardware for software development purposes. Application programs could be executed on a software simulation. This would permit trade-offs between the hardware and software implementations even before a single chip or board was designed! Such an approach based on this concept of Virtual Prototyping is an example of a newly emerging design methodology for systems in general, but is currently being applied to digital signal processing systems in particular [8,11].

1.5 Chapter Summary

Digital system design is a process of creating and managing multiple descriptions of systems representing distinct views and varying levels of abstraction. Historically, design environments grew around point tools for solving specific problems at each level of abstraction. Design methodologies governed the application of these tools and the flow of

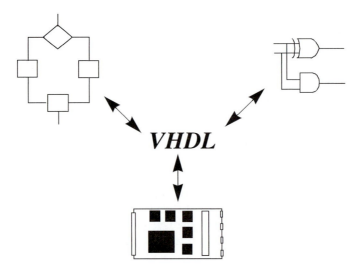

FIGURE 1-4 The VHDL language can be applied at multiple levels of abstraction

information between them. The evolution of distinct methodologies and modeling styles can hinder interoperability and make it difficult to share models. As electronic systems grew in complexity, the need to integrate these point tools into a cohesive design process has determined a large component of the economics of product design. The advent of hardware description languages such as VHDL and their acceptance as industry standards has had an enormous impact on the economics of product development.

We see that VHDL modeling can be used to model hardware and software systems at multiple levels of abstraction. The language is independent of technology and design methodologies or styles, and therefore promotes portable descriptions, rapid prototyping, and free exchange of models among organizations and individuals. The result has been the promise of reduced design cycle times, faster time to market and reduced cost. We can expect to see the use and growth of the language continue as a widely used hardware description language for both military and commercial systems for some time to come.

Within the purview of the design of electronic systems the VHDL language is accompanied by several other hardware description languages. One of the more widely used is Verilog. The goals and motivation for the Verilog language parallel those of VHDL although Verilog shares a distinct developmental heritage. The major CAD tool vendors support both the VHDL and Verilog languages.

CHAPTER 2 — Simulation of Digital Systems

VHDL programs are unlike programs written in Pascal, C, or Fortran. Conventional programs are based on thinking in terms of algorithms and sequences of calculations that manipulate data toward a specific computational goal. The thought process that goes into writing such programs is inherently procedural, a direct result of the serial nature of most modern computers. Writing VHDL programs is very different. We are not interested so much in the computation of a function, but in mimicking the behavior of some physical system such as a digital circuit. Therefore, the VHDL program must describe this physical system. An associated simulator uses this description and executes a simulation that "behaves like" the physical system. This chapter discusses the significant structural, physical, and behavioral characteristics of digital systems. It is these characteristics that the VHDL programs will describe.

Physical systems such as digital circuits do not necessarily behave like procedures. They are characterized by complex interactions between constituent components. In fact, many times the reason we want to simulate a physical system is because we cannot compute what we need to know. For example, if we need to know the average time a person must wait at a bus stop, we cannot obtain some mathematical function that will compute this value for us. Traffic patterns, whimsical pedestrians, and riders with incorrect change, all contribute to a degree of unpredictability and complexity that prevents us from writing a mathematical expression for the waiting time at a bus stop. However, we can simulate the transit system and observe how long people wait in the simulation. If our simulation is accurate, then we will be able to predict reliably the delays that will occur in practice. The digital systems that we will consider do not exhibit probabilistic or random behavior but are comprised of many constituent subsystems and can be quite complex. In the context of the design of digital systems there are several compelling reasons for simulation. We may wish to simulate a design prior to implementation to ensure that the system meets its specification. For example, we may design a board-level product that interfaces to a camera and processes images in real time. A VHDL simulation of the board may be used to establish that the design can indeed operate fast enough to keep up with the rate at which images are being received from the camera. Given the cost of modern fabrication facilities and the increasing complexity of digital systems it has become necessary to be able to rely on accurate simulation models to design and test chips and systems prior to their construction. How can we be sure that the design will function as intended or that the design is indeed correct? The VHDL simulation serves as a basis for testing complex designs and validating the design prior to fabrication. The overall effect is that of reducing redesign, shortening the design cycle, and bringing the product to market sooner.

The features of a language for describing digital systems and the behavior of the simulation itself are quite different from procedural languages. While early simulators for digital systems have been written in C or Pascal, the developers had to provide new functions, operations, and data types to enable one to write simulation applications. The definition of the VHDL language provides a range of features in support of the simulation of digital systems.

As a result of the motivation to model digital systems, many of the language concepts and constructs can be identified with the structural, behavioral, and physical characteristics of digital systems. We learn best when we can identify with concepts with which we are already familiar. In this chapter we will review the operational characteristics of digital systems and identify several key attributes. A discrete event model of the operation of a digital system is described that can be utilized to model these attributes and therefore is used for the simulation of digital circuits described in VHDL. We can think of a VHDL program as the description of a digital system while the associated simulator will use this description to produce behavior that will mimic that of the physical system.

2.1 Describing Systems

The term *system* is used in many different contexts to refer to anything from single chips to large supercomputers. Webster's dictionary defines a system as: "an assemblage of objects united by some form of regular interaction or interdependence."

We are interested in being able to describe digital systems at any one of several levels of abstraction starting from the switched transistor level to the computing system level. To do so requires us to identify attributes of systems common to all of these levels of abstraction. For example, imagine that you are in the business of selling sound cards for personal computers. You are trying to make a sale to Personal Computer, Inc., and have them include this sound card as a part of their product line making it available in all of their personal computers. A sample design of your card might appear as shown in Figure 2-1. In order to make this sale, you must be able to describe this card to the engineers of Personal Computer, Inc., How could you describe such a card? What do the engineers

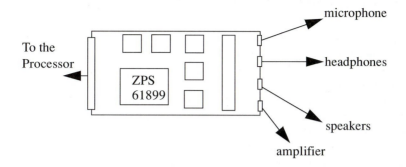

FIGURE 2-1 An example of a system

need to know to evaluate your design? They certainly need to understand the *interface* to the card. For example, what can you connect to this card? Speakers, microphones, or even a stereo amplifier? How does the processor communicate with this card? You must be able to describe all of the signals that may pass through the card interface. The second component of this description is the behavior of the card itself. This could be communicated in one of several ways. One way is to describe component chips and their interconnection assuming that the engineers were familiar with the operation of the individual chips. Such a description is commonly referred to as a *structural* description and can be easily conveyed in a block diagram. Alternatively, we can describe the behavior of the card in terms of the type of processing it performs on the input signals and the type of output signals it produces, for example, audio output for the speakers. Such descriptions are referred to as *behavioral* descriptions. You are describing what the card does independent of the physical parts that make up the card. Depending upon who you are talking to, one description or the other is preferable. For example, marketing would be interested in the latter description and engineering would be interested in the former.

Structure and behavior are complementary ways of describing systems. The specification of the behavior does not necessarily tell you anything about the structure of the system or the components used to build it. In fact, there are usually many different ways in which you can build a system to provide the same behavior. In this case, other factors such as cost or reliability become the determining factors in choosing the best design. We would expect that any language for describing digital systems will support both structural and behavioral descriptions. We would also expect that the language would enable us to evaluate or simulate several structural realizations of the same behavioral description. The VHDL language provides these features.

2.2 Events, Propagation Delays, and Concurrency

Let us look a little closer at these structural and behavioral descriptions. Digital systems are fundamentally about *signals*, specifically binary signals that may take values 0 or 1 (a more powerful value system will be introduced later). Digital circuits are comprised of *components* such as gates, flip-flops, and counters. Components are interconnected by wires and transform input signals into output signals. A machine-readable (i.e., programming language) description of a digital circuit must be able to describe the components that make up the circuit, their interconnection, and the behavior of each of the components in terms of their input and output signals and the relation between them. This language description can then be simulated by associated computer aided design tools.

Consider the gate level description of a half adder shown in Figure 2-2. There are two input signals, a, and b. The circuit computes the value of two output signals, sum and carry. The values of the output signals, sum and carry, are computed as a function of the input signals, a and b. For example, when a = 1 and b = 0, we have sum = 1 and carry = 0. Now, suppose the value of b changes to 1. We say that an *event* occurs on signal b. The event is the change in the value of the input signal from 0 to 1. In our idealized model of the world this transition takes place instantaneously at a specific, or discrete, point in time. Real circuits take a finite amount of time to switch states, but this approximation is still very useful. From the truth tables for the gates, we know that such an event on b will cause the values of the output signals to change. The signals sum and carry will acquire values 0 and 1 respectively, that is, events will occur on the signals sum and carry. A basic question then becomes the following: When will these events on the output signals occur relative to the timing of the events on the input signals?

Electrical circuits have a certain amount of inertia or natural resistance to change. Physical devices such as transistors that are used to implement the gate level logic take a finite amount of time to switch between logic levels. Therefore, a change in the value of a signal on the input to a gate will not produce an immediate change in the value of the output signal. Rather, it will take a finite amount of time for changes in the inputs to a gate to propagate to the output. This period of time is referred to as the *propagation delay*. The time it takes for changes to propagate through the gates is a function of the physical properties of the gate, including the implementation technology, the design of the gate from basic transistors, and the power supplied to the circuit. From the timing behavior depicted in Figure 2-2, we can see that the gate propagation delay is 5 ns. Electrical currents that carry logic signals through interconnect media such as the wires also travel at a finite rate. Thus, in reality, signals experience propagation delays through wires and the magnitude of this delay is dependent upon the length of the wire. This delay is non-negligible particularly in very high speed, high density circuits. It is interesting to note that wires have considerably less inertia than gates. The resulting physical phenomenon is therefore quite different. However, as device feature sizes have become increasingly smaller, wire delays have become non-negligible in modern high density circuits. As we shall see in later chapters, VHDL provides specific constructs for handling both types of delays. The timing diagram shown in Figure 2-2 does not include wire delays.

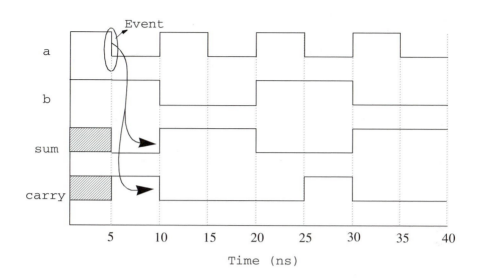

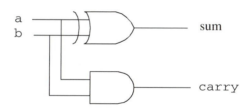

FIGURE 2-2 Half-adder circuit

A third property of the behavior of the circuit shown in Figure 2-2 is *concurrency* of operation. Once a change is observed on input signal b, the two gates concurrently compute the values of the output signals sum and carry and new events may subsequently occur on these signals. If both gates exhibit the same propagation delays then the new events on sum and carry will occur simultaneously. These new events may go on to initiate the computation of other events in other parts of the circuit. For example, consider two half adders combined to form a full adder shown in Figure 2-3. Events on the input signals In1 or In2 produce events on signals s1 or s3. Events on s1 or s3 in turn may produce events on s2, sum, or c_out. In effect, events on the input signals In1 or In2 propagate to the outputs of the full adder. In the process many other events internal to the circuit may be generated. In the associated timing diagram, every $0 \rightarrow 1$ and $1 \rightarrow 0$ transition on each signal corresponds to an event . Note the data-driven nature of these systems. Events on signals lead to computations that may generate events on other signals.

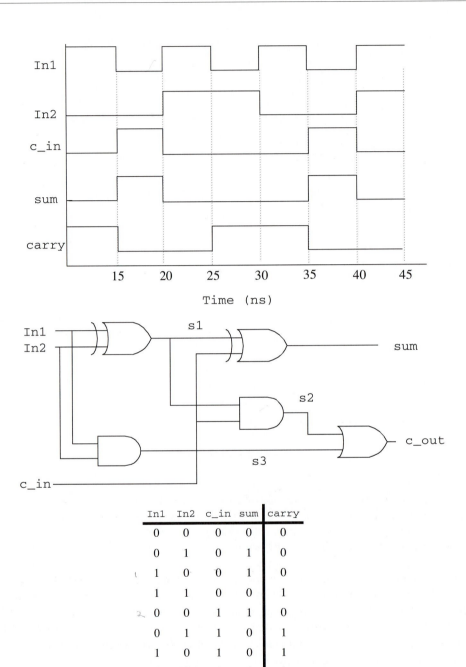

155 delay 2.

FIGURE 2-3 Full-adder circuit and truth table

In1	In2	c_in	sum	carry
0	0	0	0	0
0	1	0	1	0
1	0	0	1	0
1	1	0	0	1
0	0	1	1	0
0	1	1	0	1
1	0	1	0	1
1	1	1	1	1

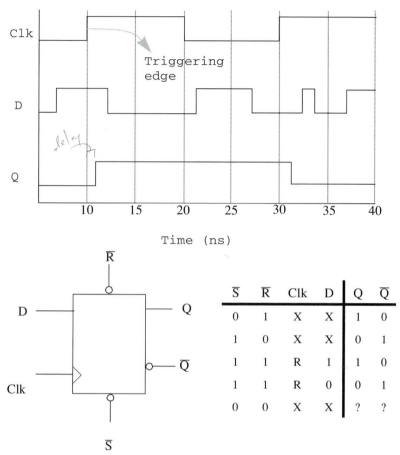

FIGURE 2-4 Excitation table for a positive edge-triggered D flip-flop
(R = rising edge and X = don't care)

2.3 Waveforms and Timing

Over a period of time, the sequence of events that occur on a signal produces a *waveform* on that signal. The effects of each event may in turn propagate through the circuit, producing waveforms on internal signals and eventually producing waveforms on the output signals. The timing diagram shown in Figure 2-3 is a collection of waveforms on signals in the full-adder circuit, where each waveform is an alternating sequence of $0 \rightarrow 1$ and $1 \rightarrow 0$ transitions or events.

The model of the operation of digital circuits in terms of events, delays, concurrent operation, and waveforms extends to sequential circuits as well as combinational circuits. Consider the operation and timing of a positive edge-triggered D flip-flop shown in Figure

2-4. The output values are determined at the time of a $0 \rightarrow 1$ transition on the clock signal. At this time the input value on signal D is sampled and the values of Q and $\overline{Q}$ are determined. Events on the set ($\overline{S}$) and reset ($\overline{R}$) lines produce events on the output independent of events on the clock. The unique aspect of the behavior of this model is the dependency on the clock signal. Computation of output events is initiated at a specific point in time determined by a $0 \rightarrow 1$ event on the Clk signal, independent of events occurring on the D input signal. This need to *wait for* a specific event is an important aspect of the behavior of sequential digital circuits. Such circuits are referred to as *synchronous* circuits. Synchronous circuits operate with a periodic signal commonly referred to as a clock that serves as a common time base. Clocks are an important aspect of digital circuits and deserve special attention

Alternatively, many digital systems operate asynchronously with request–acknowledge protocols. Operations of these systems are also characterized by the need to *wait* for specific events such as a request line being asserted.

2.4 Signal Values

Signal values are normally associated with the outputs of gates. Wires transfer these values to the inputs of other gates, which, as a result, may drive their outputs to new values. In general, we tend to think of signals in digital circuits as being binary valued and being driven to these values by a source such as the power supply or the output of a gate. These logical values are physically realized within a circuit by associating logical 0 or 1 values to voltage or current levels at the output of a device. For example, in some circuits a voltage occurring between 0 and 0.8 volts is recognized as a logical 0 signal, while a voltage occurring in the range 2.0 to 5.0 volts is recognized as logical 1 signal. However, what happens when a signal is not driven to any value, for example, if it is disconnected? What is the value of the signal? It is neither 0 or 1. Such a state is referred to as the high impedance state and is usually denoted by Z. This is a normal, inactive condition and occurs when a signal is (temporarily) disconnected.

What happens when a signal is concurrently driven to both a 0 and a 1 value? This is clearly an abnormal or error condition and should not occur. It is indicative of a design error. How do we denote the value of the signal? Remember that our overall goal is the accurate description and simulation of digital systems often for the purpose of testing a design to ensure that it is correct. If this condition were to occur during the simulation of a circuit the simulator must be able to represent the value of the signal and propagate the effects of this design error through the circuit. Such unknown values are typically denoted by X. What if the initial value of a signal is undefined? How can we represent this value and propagate the effects of uninitialized signal values to determine the effect on the operation of the circuit? Such values are typically denoted by U.

At the very least, we see that 0 and 1 values alone are insufficient to accurately capture the behavior of digital systems. We will see that the VHDL language is flexible enough to enable the definition of a range of values for signals. Early in the evolution of

VHDL, CAD tool vendors defined their own value systems. Some vendors even had as many as 46 distinct values for a binary signal! This made it difficult to share VHDL models. Imagine if different C compilers had different definitions of the values of integers. The same C program could produce different results depending upon the compilers that were used. In addition to 0, 1, Z, X, and U values, it is useful to denote the concept of *signal strength.* The strength of a signal reflects the ability of the source device to supply energy to drive the signal. This strength can be weakened or attenuated by, for example, the resistance of the wires giving rise to signals of different strengths. Although the range of strength values can be large, only two levels are sufficient to characterize certain types of transistor circuits. The use of strength values facilitates certain styles of design. The VHDL language is being widely used to describe the behavior of circuits that can be automatically synthesized by design tools. A value system that incorporates the concept of signal strength is therefore necessary.

Value	Interpretation
U	Uninitialized
X	Forcing Unknown
0	Forcing 0
1	Forcing 1
Z	High Impedance
W	Weak Unknown
L	Weak 0
H	Weak 1
–	Don't Care

FIGURE 2-5 IEEE 1164 Value System

In an attempt to establish common ground and enable the construction of portable models, the IEEE has approved a 9-value system. This is the IEEE 1164 standard, which is rapidly gaining acceptance and widespread usage. In this system binary signals take on functional values of 0, 1. However, they can also be unknown (X), uninitialized (U), or not driven (Z). If we include two levels of signal strength and the don't-care value (–), we have the IEEE 1164 value system shown in Figure 2-5. It is important to note that this value system is not a part of the VHDL language but rather a standard definition that vendors are motivated to support and users are motivated to use since it enables reuse of designs and sharing of models between users. Practically all vendors support the IEEE 1164 value system.

2.5 Shared Signals

It is common for components in a digital circuit to have multiple sources for the value of an input signal. However, connecting all sources and destinations by dedicated signal paths can be expensive. Therefore many designs will utilizes *buses*: a group of signals that can be shared among multiple source and destination components. For example, the architecture of personal computers and workstations is built around one or more buses. The microprocessor chip may drive a data bus to communicate values to memory, while, at other times, the memory controller may drive the bus to return values. The Input/Output buses in PCs interconnect many devices such as CD ROMs, floppy disk drives, and hard disk drives, that can be the source of values on the Input/Output bus, although not at the same time. Certain forms of switching circuits are carefully designed based on *wired logic*. In these circuits, the interconnection of wires can produce AND and OR Boolean functions. For example, if several devices drive a shared signal to either 0, or Z, then the signal value will be determined by the interactions among all of the values applied it: in this case if at least one device is driving the signal to a 0, the value of the signal will be 0. Thus, the interconnection can be regarded as implementing a wired-OR function.

In general, when multiple drivers exist for a signal, what value does the signal have? Clearly, this depends on the implementation. Shared signals such as buses are supposed to have only one active source at any given time, but this is not the case for circuits based on wired logic. A hardware description language must be capable of describing the interactions between multiple drivers for shared signals.

2.6 A Discrete Event Simulation Model

The preceding sections described the behavior of digital systems in terms of events that take place at discrete points in time. Some events may cause other events to be generated after some delay and many events may be generated concurrently. *Discrete event simulation* is a programming-based methodology for accurately modeling the generation of events in physical systems. The operation of a physical system such as a digital circuit is described in a computer program that specifies how and when events—changes in signal values—are generated. A *discrete event simulator* then executes this program, modeling the passage of time and the occurrence of events at various points in time. Such simulators often manage millions of events and rely on well-developed techniques to keep track accurately of the *correct order* in which the events occur. We can view VHDL as a programming language for describing the generation of events in digital systems supported by a discrete event simulator. The following describes a simple discrete event simulation model that captures the basic elements of the simulation of VHDL programs. Understanding the VHDL model of time is a necessary prelude to writing, debugging, and understanding VHDL models.

Discrete event simulations utilize an event list data structure. The event list maintains an ordered list of all future events in the circuit. Each event is described by the type of event—a $0 \rightarrow 1$ or $1 \rightarrow 0$ transition—and the time at which it is to occur. Although we generally think of transitions as a change of value between 0 and 1, recall that signals may have other values. In general, an event is simply a change in the value of a signal. Let us refer to the time at which an event is to occur as the *timestamp* of that event. The event list is ordered according to increasing timestamp value. This enables the simulator to execute events in the order that they occur in the real world, that is, the physical system. Finally, the simulator clock records the passage of simulated time. The value of this clock will be referred to as the current *timestep*, or simply timestep. Imagine what would happen if we froze the system at a timestep and took a snapshot of the values of all of the signals in the system. These values would represent the state of the simulation at that point in time.

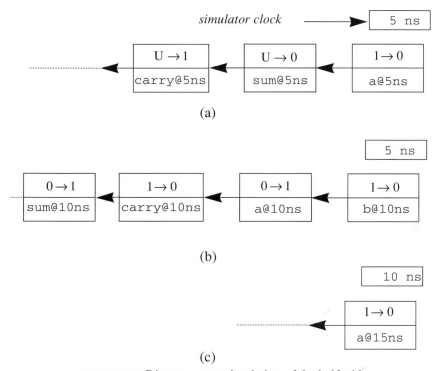

FIGURE 2-6 Discrete event simulation of the half adder

Let us consider one approach to the discrete event simulation of the half-adder circuit shown in Figure 2-2. Assume that we have been able to specify waveforms on the
inputs a and b. In a physical circuit these waveforms would likely be generated as the output of another component. At timestep 0 ns, initial values on inputs a and b will cause
events that set the values of the sum and carry signals. Assuming that the propagation
delay of a gate is 5 ns, these events will be scheduled 5 ns later. Figure 2-6(a) shows these
events at the head of the event list at time 5 ns. Note that both sum and carry are scheduled to receive values at this time. Prior to this time the values of the sum and carry are
undefined, as represented by the shaded areas in the timing diagram shown in Figure 2-2.
Input a is also scheduled to make a transition at the same time (for the moment let us
ignore how these events on the input signal are generated). These events correspond to the
signal transitions shown on the timing diagram in Figure 2-2 at 5 ns. The simulator
removes these events from the event list, and the current values of these signals are
updated. Due to a change in the value of signal a, the simulator determines that the values
of the output signals, sum and carry, have to be recomputed. The computation produces
new values of sum and carry, which are scheduled in the event list at timestep 10 ns.
The head of the event list now appears as shown in Figure 2-6(b), with all of the events
scheduled for timestep 10 ns. The global clock is now updated to 10 ns, all events sched-

uled at timestep 10 ns are removed from the event list, the corresponding signal values are updated, and any new events are computed and scheduled. Figure 2-6(c) shows the head of the event list after event computations at timestep 10 ns, and prior to the update of the global clock. The simulator clock will now be updated to 15 ns and the process repeated. This process is continued until there are no more events to be computed or until some predetermined simulation time has expired. This preceding behavior of a discrete event simulator can be described in the following steps:

1. Advance simulation time to that of the event with the smallest timestamp in the event list. This is the event at the head of the list.
2. Execute all events at this timestep by updating signal values.
3. Execute the simulation models of all components affected by the new signal values.
4. Schedule any future events.
5. Repeat until the event list is empty, or a preset simulation time has expired.

In general, there is substantial concurrency in a digital circuit and many events may take place simultaneously. Thus, many signals may receive values at the same timestep and more than one event is executed at a timestep. We see that the simulator employs a two-stage model of the evolution of time. In the first stage, simulation time is advanced to that of the next event, and all signals receiving values at this time are updated. In the second stage, all components affected by these signal updates are reevaluated, and any future events that are generated by these evaluations are scheduled by placing them into the event list in order of their timestamp.

It is apparent that this model is quite flexible and general. We can think of modeling gate level circuits as well as higher level circuits such as arithmetic logic units (ALUs), decoders, multiplexors, and even microprocessors. We simply need to describe the behavior of these components in terms of input events, computation of the output events from input events, and propagation delays. If we can describe a digital system in these terms, we can develop computer programs for implementing this behavior. Note that discrete event simulation itself is only a model of the behavior of real systems. In real circuits signals do not make instantaneous transitions between logic 0 voltage levels and logic 1 voltage levels. For that matter there are no such things as truly digital devices. There are only analog devices wherein we interpret analog voltage levels as 0 or 1! For many purposes, such a discrete event model is adequate. However, often more detailed and accurate models are required, in which case complementary techniques and models are employed. To distinguish the discrete event model from the real system, we will refer to the former as the *logical model* and the latter as the *physical system*.

2.7 Chapter Summary

This chapter is based on the premise that we must understand the execution model of the language before we can use it effectively. The VHDL language was motivated by the need to accurately model and simulate digital systems. This chapter has focused on the attributes of the behavior of digital systems. By identifying these attributes in this chapter and establishing the terminology, it was possible to describe a general discrete event model for the execution of VHDL programs. I hope that this will provide an intuitive basis for the reader to learn and apply the VHDL language constructs described in succeeding chapters.   The key attributes discussed in this chapter include:

- System descriptions
    - structural
    - behavioral
- Events
- Propagation delays
- Concurrency
- Timing
    - synchronous
    - asynchronous
- Waveforms
- Signal values
- Shared signals
- Discrete event simulation

   The VHDL language provides basic constructs for representing each of the above attributes. The associated simulator implements a discrete event simulation model, manages the progression of simulated time, and maintains internal representations of the waveforms being generated on signals. The language constructs to specify these attributes are described in the following chapters.

CHAPTER 3

Basic Language Concepts

VHDL has often been criticized as being overly complex and intimidating to the novice user. While the language is extensive, a quick start towards building useful simulation models can be made by relying on a core set of language constructs. This chapter describes the language constructs provided within VHDL for describing the attributes of digital systems identified in Chapter 2, such as, events, propagation delays, concurrency, and waveforms. Chapter 4 then introduces concepts that extend the constructs introduced here to enable the application of conventional programming constructs in building models of complex digital systems, particularly at higher levels of abstraction. Collectively these two chapters provide us with the tools necessary to model all of the attributes of digital systems described in Chapter 2.

3.1 Signals

Conventional programming languages manipulate basic objects such as variables and constants. Variables receive values through assignment statements and can be assigned new values through the course of a computation. Constants, on the other hand, may not change their values. In contrast, digital systems are fundamentally about *signals*. We have seen that signals may take on one of several values such as 0, 1, or Z. Signals are analogous to the wires used to connect components of a digital circuit. To capture the behavior of digital signals the VHDL language introduces a new type of programming object: the `signal` object type. Like variables, signals may also be assigned values, but differ from variables in that they have an associated *time value,* since a signal receives a value at a specific point in time. The signal retains this value until it is assigned a new value at a future point in time. The sequence of values assigned to a signal over time is the *waveform* of the signal. It is primarily this association with time–value pairs that differentiates a signal from a variable. A variable always has one current value. At any instant in time a signal may be associated with several time–value pairs, where each time–value pair represents some future value of the signal. Finally, we note that variables may be declared to be of a specific type, such as **integer**, **real**, or **character**. In a similar manner, a signal can be declared to be of a specific type. When used in this way, a signal does not necessarily have correspondence with the wires that connect digital components. For example, we may model the output of an ALU as an integer value. This output is treated as a signal and in simulation behaves as a signal by receiving values at specific points in time. However, we do not have to concern ourselves with modeling the number of bits necessary at the output of the ALU. This behavior enables us to model systems at a higher level of abstraction than digital circuits. Such high-level simulation is useful in the early stages of the design process, where many details of the design are still being developed.

Before we can understand how to declare and operate on signals we must first cover the basic programming constructs in VHDL. We will return to discuss signal objects in greater detail later in this chapter.

3.2 Entity–Architecture

We start by addressing the issue of describing digital systems. The primary programming abstraction in VHDL is a *design entity*. Examples of design entities include a chip, board, and transistor. It is a component of a design whose behavior is to be described and simulated. Consider once again the gate level digital circuit for a half adder shown in Figure 3-1. There are two input signals, a, and b. The circuit computes the value of two output signals, sum and carry. This half-adder circuit represents an example of a design entity.

How can the half-adder circuit be accurately described? Imagine that you had to describe this circuit over the telephone to a friend who was familiar with digital logic gates but was not familiar with the half adder. Your description would most likely include the input signals, the output signals, and a description of the behavior. The behavior in turn may be specified with a truth table, Boolean equations, or simply an interconnection between gates. We observe that there are two basic components to the description of any design entity, (i) the interface to the design, and (ii) the internal behavior of the design. The VHDL language provides two distinct constructs to specify the interface and internal behavior of design entities, respectively.

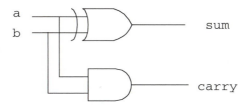

FIGURE 3-1 Half-adder circuit

The external interface to this entity is specified with the **entity** declaration. For the circuit shown in Figure 3-1, the entity declaration would appear as follows:

> **entity** half_adder **is**
> **port**(a, b : **in bit**;
>   sum, carry: **out bit**);
> **end** half_adder;

The boldface type denotes keywords that are VHDL reserved keywords. The remaining are user supplied. Just as we name programs, the label half_adder is the name given to this design entity by the programmer. The VHDL language is *case insensitive*. The inputs and outputs of the circuit are referred to as *ports*. The ports are special programming objects and are signals. Ports are the means by which the half adder can communicate with the external world or other circuits. Therefore, naturally, we expect ports to be signals rather than variables. Like variables in conventional programming languages, each port must be a signal that is declared to be of a specific type. In this case each port is declared to be of type **bit**, and represents a single-bit signal. A **bit** is a signal type that is defined within the VHDL language, and can take the values of 0 or 1. A **bit_vector** is a signal type comprised of a vector of signals, each of type **bit**. The type **bit** and **bit_vector** are two common types of ports. In general, a port may be one of several other VHDL data types. Common data types and operators supported by the language are described in Chapter 9.

From our study of digital logic, we know that bits and bit vectors are fundamental signals. From Chapter 2, we know that in reality signals can take on many different values other than 0 and 1. In practice, the types **bit** and **bit_vector** are of limited use. The IEEE 1164 Standard is gaining widespread acceptance as a value system. In this standard the 9-value signal would be declared to be of type std_ulogic rather than **bit**. Analogously, we would have the type std_ulogic_vector rather than **bit_vector**. Therefore, throughout the remainder of this text, all examples will utilize the IEEE 1164

standard signal and data types. The preceding entity declaration would now appear as follows.

entity half_adder **is**
port(a, b: **in** std_ulogic;
        sum, carry: **out** std_ulogic);
end half_adder;

The signals appearing in a port declaration may be distinguished as input signals,
output signals, or bidirectional signals. This is referred to as the *mode* of the signal. In the
above example, the **in** and **out** specifications denote the mode of the signal. Bidirectional
signals are of mode **inout**. Every port in the entity description must have its mode and type
specified.

We see that it is relatively straightforward to write the entity descriptions of standard
digital logic components. The following shows some sample circuits and their entity
descriptions. Note how byte and word-wide groups of bits are specified. For example a 32
bit quantity is declared to be of the type std_ulogic_vector (31 **downto** 0). This
type refers to a data item that is 32 bits long where bit 31 is the most significant bit in the
word and bit 0 is the least significant bit in the word.

Example: Entity Declaration of a 4-to-1 Multiplexor

entity mux **is**
port (I0, I1 : **in** std_ulogic_vector (7 **downto** 0);
        I2, I3 : **in** std_ulogic_vector (7 **downto** 0);
        Sel   : **in** std_ulogic_vector (1 **downto** 0);
        Z   : **out** std_ulogic_vector (7 **downto** 0));
end mux;

Example End: Entity Declaration of a 4-to-1 Multiplexor

Example: Entity Declaration of a D Flip-Flop

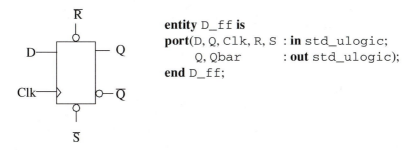

```
entity D_ff is
port(D, Q, Clk, R, S : in std_ulogic;
     Q, Qbar        : out std_ulogic);
end D_ff;
```

Example End: Entity Declaration of a D Flip-Flop

Example: Entity Declaration of a 32-bit ALU

```
entity ALU32 is
port(A, B : in std_ulogic_vector (31 downto 0);
     C    : out std_ulogic_vector (31 downto 0);
     Op   : in std_ulogic_vector (5 downto 0);
     N, Z : out std_ulogic);
end ALU32;
```

Example End: Entity Declaration of a 32-bit ALU

From the preceding examples it is clear that design entities can occur at multiple levels of abstraction, from the gate level to large systems. In fact, it should be apparent that a design entity does not even have to represent digital hardware! The description of the interface is simply a specification of the input and output signals of the design entity.

Once the interface to the digital component or circuit has been described, it is now necessary to describe its internal behavior. The VHDL construct that enables us to specify the behavior of a design entity is the **architecture** construct. The syntax of the architecture construct is the following.

> **architecture** behavioral **of** half_adder **is**
> *-- place declarations here*
> **begin**
>   *-- place description of behavior here --*
> **end** behavioral;

The above construct provides for the declaration of the module named behav-ioral which will contain the description of the behavior of the design entity named half_adder. Such a module is referred to as the **architecture** and is associated with the entity named in the declaration. Thus, the description of a design entity takes the form of an entity–architecture pair. The architecture description is linked to the correct entity description by providing the name of the corresponding entity in the first line of the architecture.

The behavioral description provided in the architecture can take many forms. These forms differ in the levels of detail, description of events, and the degree of concurrency. The remainder of this chapter focuses on a core set of language constructs required to model the attributes of digital systems described in Chapter 2. Subsequent chapters will add constructs motivated by the need for expanding the scope and level of abstraction of the systems to be modeled.

3.3   Concurrent Statements

The operation of digital systems is inherently concurrent. Many components of a circuit can be simultaneously operating and concurrently driving distinct signals to new values. How can we describe the assignment of values to signals? We know that signal values are time–value pairs, that is, a signal is assigned a value at a specific point in time. Within VHDL signals are assigned values using *signal assignment* statements. These statements specify a new value of a signal and the time at which the signal is to acquire this value. Multiple signal assignment statements are executed concurrently in simulated time and are referred to as *concurrent signal assignment statements (CSAs)*. There are several forms of CSA statements and they are described in the following section.

3.3.1 Simple Concurrent Signal Assignment

Consider a description of the behavior of the half-adder circuit shown in Figure 3-1. Recall that although VHDL manages the progression of time, we need to be able to specify events, delays, and concurrency of operation.

```
architecture concurrent_behavior of half_adder is
begin
 sum <= (a xor b) after 5 ns;
 carry <= (a and b) after 5 ns;
end concurrent_behavior;
```

Just as we named entity descriptions, the label `concurrent_behavior` is the name given to this architecture module. The first line denotes the name of the entity that contains the description of the interface of this design entity. Each statement in the above architecture is a *signal assignment* statement with the operator "<=" denoting signal assignment. Each statement describes how the value of the output signal depends on, and is computed from, the value of the input signals. For example, the value of the sum output signal is computed as the Boolean exclusive-OR operation of the two input signals. Once the value of sum has been computed, it will not change unless the value of a or b changes. Figure 3-2 illustrates this behavior. At the current time, a = 0, b = 1, and sum = 1. At time 10, the value of b changes to 0. The new value of the sum will be (a **xor** b) = 0. Since there will be a propagation delay through the exclusive-OR gate, the signal sum will be assigned this value 5 ns later at time 15. This behavior is captured in the first signal assignment statement. Note that, unlike variable assignment statements, the signal assignments shown above specify both value and (relative) time.

In general, if an event (signal transition) occurs on a signal on the right-hand side of a signal assignment statement, the expression is evaluated and new values for the output signal are scheduled for some time in the future as defined by the **after** keyword. The dependency of the output signals on the input signals is captured in the two statements and **NOT** in the textual order of the program. The order of the statements could be reversed and the behavior of the circuit would not change. Both statements are executed concurrently with respect to simulated time to reflect the concurrency of corresponding operations in the physical system. This is why these statements are referred to as concurrent signal assignment statements. A fundamental difference between VHDL programs and conventional programming languages is that concurrency is a natural part of the systems described in VHDL and therefore of the language itself. Note that the execution of the statements is determined by the flow of signal values, rather than textual order. Figure 3-2 shows a complete, executable half-adder description and the associated timing behavior. This description contains the most common elements used to describe a design entity.

Note the use of the **library** and **use** clauses. We can think of libraries as repositories for frequently used design entities that we wish to share. The **library** clause identifies a library that we wish to access. The name is a logical name for a library. In Figure 3-2 the library name is IEEE. In practice, this logical name will usually map to a directory on the local system. This directory will contain various design units that have been compiled. A *package* is one such design unit. A package may contain definitions of types, functions, or procedures to be shared by multiple application developers. The **use** clause determines which of the packages or other design units in a library you will be using in the current design. In the preceding example, the use clause states that in library IEEE there is a

```
library IEEE;
use IEEE.std_logic_1164.all;

entity half_adder is
port (a, b : in std_ulogic;
      sum, carry : out std_ulogic);
end half_adder;

architecture concurrent_behavior of half_adder is
begin
sum <= (a xor b) after 5 ns;
carry <= (a and b) after 5 ns;
end concurrent_behavior;
```

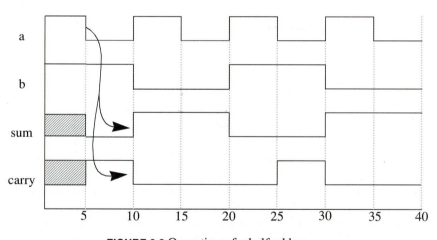

FIGURE 3-2 Operation of a half adder

package named std_logic_1164, and that we will be able to use all of the components defined in this package. We need this package, since the definition for the type std_ulogic is in this package. The VHDL models that use the IEEE 1164 value system will include the package declaration as shown. Design tool vendors typically provide the library IEEE and the std_logic_1164 package. These concepts are analogous to the use of libraries for mathematical functions and input–output in conventional programming languages. Libraries and packages are described in greater detail in Chapter 6. This example now contains the major components found in VHDL models: declarations of existing design units in libraries that you will be using, entity description of the design unit, and the architecture description of the design unit.

The descriptions provided so far in this chapter are based on the specification of the value of the output signals as a function of the input signals. In larger and more complex

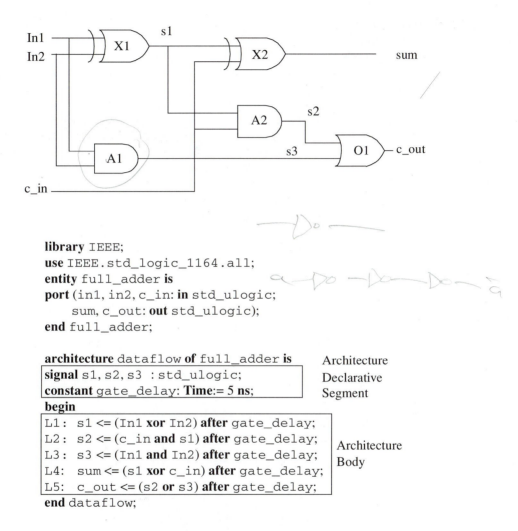

```
library IEEE;
use IEEE.std_logic_1164.all;
entity full_adder is
port (in1, in2, c_in: in std_ulogic;
      sum, c_out: out std_ulogic);
end full_adder;
```

```
architecture dataflow of full_adder is          Architecture
signal s1, s2, s3 : std_ulogic;                 Declarative
constant gate_delay: Time:= 5 ns;               Segment
begin
L1:  s1 <= (In1 xor In2) after gate_delay;
L2:  s2 <= (c_in and s1) after gate_delay;      Architecture
L3:  s3 <= (In1 and In2) after gate_delay;      Body
L4:  sum <= (s1 xor c_in) after gate_delay;
L5:  c_out <= (s2 or s3) after gate_delay;
end dataflow;
```

FIGURE 3-3 VHDL model of a full adder

designs, there are usually many internal signals used to connect design components such as gates or other hardware building blocks. The values that these signals acquire can also be written using simple concurrent signal assignment statements. However, we must be able to declare and make use of signals other than those within the entity description. The gate level description of the full adder provides an example of such a VHDL model.

Example: Full-Adder Model

Consider the full-adder circuit shown in Figure 3-3. We are interested in an accurate simulation of this circuit where all of the signal transitions in the physical realization are modeled. In addition to the ports in the entity description, we see that there are three internal signals. These signals are named and declared in the architectural description. The declarative region declares three single-bit signals: s1, s2, and s3. These signals are annotated in the circuit. Now we are ready to describe the behavior of the full adder in terms of the internal signals as well as the entity ports. Since this circuit uses two input gates, each signal is computed as a Boolean function of two other signals. The model is a simple statement of *how* each signal is computed as a function of other signals, and the propagation delay through the gate. There are two output signals and three internal signals, for a total of five signals. Accordingly, the description consists of five concurrent signal assignment statements, one for each signal.

   Each signal assignment statement is given a label: L1, L2, and so on. This labeling is optional, and can be used for reference purposes. Note a new language feature in this model—the use of the **constant** object. Constants in VHDL function in a manner similar to conventional programming languages. A constant can be declared to be of a specific type, in this case of type **Time**. A constant must have a value at the start of the simulation and cannot be changed during the simulation. At this stage, it is easiest to ensure that constants are initialized as shown above. The introduction of the type **Time** is a natural consequence of simulation modeling. Any object of this type must take on the values of time such as microseconds or nanoseconds. The type **Time** is a predefined type of the language.

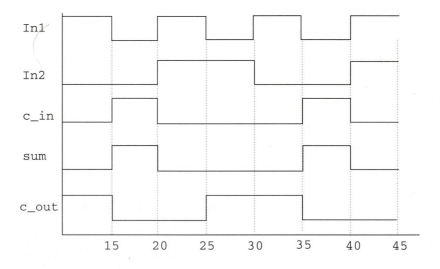

FIGURE 3-4 Full-adder circuit timing

As we know, the textual order of the statements is irrelevant to correct operation of the circuit model. Let us now consider the flow of signal values and the sequence of execution of the signal assignment statements. Figure 3-4 shows the waveforms of all of the signals in the full-adder circuit. From the figure we see that there is an event on In1 at time 10 changing its value to 1. This causes statements L1 and L3 (from Figure 3-3) to be executed and new values to be scheduled on signals s1 and s3 at time 15. These events in turn cause statements L2 and L5 to be executed at time 20 and events to be scheduled on signals c_out and s2 at time 20. We see that the execution of the statement L1 produced events that caused the execution of statement L5. This order of execution is maintained regardless of the textual order in which they appear in the program.

Note the two-stage model of execution. In the first stage all statements with events occurring at the current time on signals on the right hand side (RHS) of the signal assignment statement are evaluated. All future events that are generated from the execution of these statements are then scheduled. Time is now advanced to the time of the next event. The process is repeated. Note how the programmer specifies events, delays, and concurrency. Events are specified with signal assignment statements. Delays are specified within the signal assignment statement. Concurrency is specified by having a distinct signal assignment statement for each signal. The order of execution of the statements is dependent upon the flow of values (just as the case in the real circuit) and not on the textual order of the program. As long as the programmer correctly specifies how the value of each signal is computed and when it acquires this value relative to the current time, the simulator will correctly reflect the behavior of the circuit.

Example End: Full-Adder Model

3.3.2 Implementation of Signals

Unlike variables, signals are a new type of programming object and merit specific attention. So far we have seen that signals can be declared in the body of an architecture or in the port declaration of an entity. The form of the declaration is

signal s1 : std_ulogic := '0';

or, more generally,

identifier-list : type := expression;

If the signal declaration included the assignment symbol (i.e.,:=) followed by an expression, the value of the expression is the initial value of the signal. The initialization is not required, in which case the signal is assigned a default value as specified by the type definition. Signals can be one of many valid VHDL types: integers, real, bit_vector, and so forth.

Now consider the assignment of values to a signal. We know that signal assignment statements assign a value to a signal at a specific point in time. The simple concurrent signal assignment statements described so far in this chapter exhibit the following common structure:

> sum <= (a **xor** b) **after** 5 **ns**;

which can be written in a more general form as

> *signal* <= *value expression* **after** *time expression;*

The expression on the right hand side of the signal assignment is referred to as a *waveform element*. A waveform element describes an assignment to a signal and is comprised of a *value expression* to the left of the **after** keyword and a *time expression* to the right of the keyword. The former evaluates to the new value to be assigned to the signal and the latter evaluates to the relative time at which the signal is to acquire this value. In this case the new value is computed as the exclusive-OR of the current values of the signals a and b. The value of the time expression is added to the current simulation time to determine when the signal will receive this new value. In this case the time expression is a constant value of 5 ns. With respect to the current simulation time, this time–value pair represents the future value of the signal and is referred to as a *transaction*. The underlying discrete event simulator that executes VHDL programs must keep track of all transactions that occur on a signal. The list is ordered in increasing time of the transactions.

If the evaluation of a single waveform element produces a single transaction on a signal, can we specify multiple waveform elements and, as a result, multiple transactions? For example, could we have the following?

> s1 <= (a **xor** b) **after** 5 **ns**, (a or b) **after** 10 **ns**, (**not** a) **after** 15 **ns**;

The answer is, yes! When an event occurs on either of the signals a or b, then the above statement would be executed, all three waveform elements would be evaluated, and three transactions would be generated. Note that these transactions are in increasing order of time. The events represented by these transactions must be scheduled at different points in the future, and the VHDL simulator must keep track of all of the transactions that are currently scheduled on a signal. This is achieved by maintaining an ordered list of all of the current transactions pending on a signal. This list is referred to as the *driver* for the signal. The current value of a signal is the value of the transaction at the head of the list. What is the physical interpretation of such a sequence of events? These events represent the value of the signal over time, which essentially is a waveform. This is how we can represent a signal waveform in VHDL: as a sequence of waveform elements. Therefore, within a signal assignment statement, rather than assigning a single value to the signal at some future time we can assign a waveform to this signal. This waveform is specified as a sequence of signal values and each value is specified with a single waveform element. Within the simulator these sequences of waveform elements are represented as a sequence of transactions on the driver of the signal. These transactions are referred to as the *projected output waveform*, since these events have not yet occurred in the simulation. What if the simulation attempts to add transactions that conflict with the current projected waveform? The VHDL language definition provides specific rules for adding transactions to the projected waveform of a signal.

Example: Specifying Waveforms

Assume that we would like to generate the following waveform. We could do so with the signal assignment statement shown below.

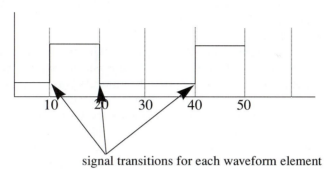

signal transitions for each waveform element

signal <= '0','1' **after** 10 **ns**,'0' **after** 20 **ns**,'1' **after** 40 **ns**;

Note how each transition in the above waveform is specified as a single waveform element in the signal assignment statement. All waveform elements must be ordered in increasing time. Failure to do so will result in an error.

Example End: Specifying Waveforms

Note how the concepts and terminology discussed so far are derived from the operation of digital circuits. In a physical circuit a wire (signal) has a driver associated with it. Over time this driver produces a waveform on that wire. If we continue to view the language constructs by analogy with the digital circuits they were intended to model, it will be easier for us to reason about the construction of models using VHDL. The constructs that manipulate signals invariably rely on waveform elements to specify input and output waveforms. Understanding this representation is key to understanding many of the VHDL programming constructs.

3.3.3 Resolved Signals

Our view of signals up to this point has been one where every signal has only one driver, that is, one signal assignment statement that is responsible for generating the waveform on that signal. We know that is not true in practice. Shared signals occur on buses and in circuits based on wired logic. When a signal has multiple drivers, how is the value of the signal determined? In the VHDL language, this value is determined by a *resolution function*.

A resolution function examines all of the drivers on a shared signal and determines the value to be assigned to the signal. A shared signal must be of a special type: a *resolved type*. A resolved type has a resolution function associated with the type. In the preceding examples, we have been using the std_ulogic and std_ulogic_vector types for single-bit and multibit signals respectively. The corresponding resolved types are std_logic and std_logic_vector. This distinction has the following consequences. In the course of the simulation, when a signal of type std_logic is assigned a value, the associated resolution function is automatically invoked to determine the correct value of the signal. Multiple drivers for this signal may be projecting multiple future values for this signal. The resolution function examines these drivers to return the correct value of the signal at the current time. If the signal has only one driver, then determination of the value is straightforward. However, if more than one driver exists for the signal the value that is assigned to the signal is the value determined by the resolution function. For the IEEE 1164 package, the resolution function is essentially a lookup table. Provided with the signal values from two drivers, the table returns the signal value to be assigned. For example, if one source is driving the signal to 1 and a second source's output is left floating (i.e., in state Z), the resulting value will be 1. Alternatively, if the two sources are driving the shared signal to 1 and 0 respectively, the resulting value will be unknown or X. The resolution function for the std_logic and std_logic_vector types is provided by the std_logic_1164 package. Having multiple drivers for a signal whose type is an unresolved type will result in an error. The user may define new resolved types and provide the resolution functions for their use.

We will leave resolution functions for the moment and return to them in greater detail when we deal with the creation and use of procedures and functions in Chapter 6. However, in the remainder of this text all of the examples use the IEEE 1164 resolved single-bit and multibit types, std_logic and std_logic_vector, rather than unresolved types std_ulogic or std_ulogic_vector.

Simulation Exercise 3.1: A First Simulation Model

This exercise introduces the construction and simulation of simple VHDL models. The following steps require simulator-specific commands. Tutorials for two simulators can be found in the Appendices.

Step 1.   Using a text editor, create a VHDL model of the full adder shown in Figure 3-3. Do not use a word processor even though they may have an option for saving your text as an ASCII file. Some word processors place control characters in the file or may handle some characters non-uniformly, for example, left and right quotation marks. This can lead to analyzer errors (however you could correct these errors at that time). Set the gate delays to 3 ns for the EX-OR gates and to 2 ns for all of the other gates.

Step 2.   Use the types std_logic and std_logic_vector for the input and output signals. Declare and reference the library IEEE and the package std_logic_1164.

Step 3. Compile and load the model for simulation using a VHDL simulator toolset.

Step 4. Generate a waveform on each of the input signals.

Step 5. Run the simulation for 40 ns and trace (i) the input signals, (ii) the internal signals, s1, s2, and s3, and (iii) the sum and carry outputs.

Step 6. Check and list scheduled events on the internal signals and output signals.

Step 7. Pick an event on one of the input signals. Record the propagation of the effect of this event through the signal trace. Study the trace and ensure that the model is operating correctly.

Step 8. Repeat this example, only this time do not initialize one of the input signals. What does the resulting trace look like and what is the significance of the values on this uninitialized input?

End Simulation Exercise 3.1

3.3.4 Conditional Signal Assignment

The simple concurrent signal assignment statements that we have seen so far compute the value of the target signal based on Boolean expressions. The values of the signals on the RHS of the signal assignment statement are used to compute the value of the target signal. This new value is scheduled at some point in the future using the **after** keyword. Expressing values of signals in this manner is convenient for describing combinational circuits whose behavior can be expressed with Boolean equations. However, we often need to model high-level circuits such as multiplexors and decoders that require a richer set of constructs.

For example, consider the physical behavior of a 4-to-1, 8-bit multiplexor shown in Figure 3-5. The value of Z is one of In0, In1, In2, or In3. The waveform that appears on one of the inputs is transferred to the output Z. The specific choice depends upon the value of the control signals S0 and S1 for which there are four possible alternatives. Each of these must be tested and one chosen. This behavior is captured in the conditional signal assignment statement and is illustrated for a 4-to-1, 8-bit multiplexor in Figure 3-5. The structure of the statement follows from the physical behavior of the circuit. For each of the four possible values of S0 and S1, a waveform is specified. In this case, the waveform is comprised of a single waveform element describing the most recent signal value on that input. As pointed out in Section 3.3.2, more than one waveform element in each line of the conditional statement could have been specified, producing a waveform on the output signal Z.

In the corresponding physical circuit, an event on any one of the input signals, In0–In3, or any of the control signals, S0 or S1, may cause a change in the value of the output signal. Therefore, whenever any such event takes place the concurrent signal assignment statement is executed and all four conditions may be checked. The order of the statements is important. The expressions in the RHS are evaluated in the order that they

```
library IEEE;
use IEEE.std_logic_1164.all;
entity mux4 is
port (In0, In1, In2, In3  : in std_logic_vector (7 downto 0);
      S0, S1: in std_logic;
      Z : out std_logic_vector (7 downto 0));
end mux4;

architecture behavioral of mux4 is
begin
Z <= In0 after 5 ns when S0 = '0' and S1 = '0' else
     In1 after 5 ns when S0 = '0' and S1 = '1' else
     In2 after 5 ns when S0 = '1' and S1 = '0' else
     In3 after 5 ns when S0 = '1' and S1 = '1' else
     "00000000" after 5 ns;
end behavioral;
```

FIGURE 3-5 Conditional signal assignment statement

appear. The first conditional expression that is found to be true determines the value that is transferred to the output. Therefore, we must be careful in ordering the conditional expressions on the RHS to reflect the order in which they would be evaluated in the corresponding physical system. A careful look at the example in Figure 3-5 will reveal that, in this case, only one expression can be true and therefore the order does not matter in this particular example. Finally, note that in Figure 3-5 even though there several lines of text this corresponds to only one signal assignment statement.

3.3.5 Selected Signal Assignment Statement

The selected signal assignment statement is very similar to the conditional signal assignment statement. The value of a signal is determined by the value of a *select expression*. For example, consider the operation of reading the value of a register from a register file with eight registers. Depending upon the value of the address, the contents of the appropriate register are selected. An example of a read-only register file with two read ports is shown in Figure 3-6.

This statement operates very much like a case statement in conventional languages. As a result, its semantics is somewhat distinct from conditional signal assignment statements. The choices are not evaluated in sequence. All choices are evaluated, but only one must be true. Furthermore, all of the choices that the programmer specifies must cover all of the possible choices. For example, consider the VHDL code shown in Figure 3-6. Assume that we have only 4 registers but both addr1 and addr2 are 3-bit addresses and therefore can address up to eight registers. The VHDL language requires you to specify

```
library IEEE;
use IEEE.std_logic_1164.all;

entity reg_file is
port (addr1, addr2: in std_logic_vector (2 downto 0);
     reg_out_1, reg_out_2: out std_logic_vector (31 downto 0));
end reg_file;

architecture behavior of reg_file is
signal  reg0, reg2, reg4, reg6: std_logic_vector (31 downto 0):=
to_stdlogicvector(x"12345678");
signal reg1, reg3,  reg5, reg7: std_logic_vector (31 downto 0):=
to_stdlogicvector(x"abcdef00");
begin
with addr1 select
reg_out_1  <= reg0 after 5 ns when "000",
                reg1 after 5 ns when "001",
                reg2 after 5 ns when "010",
                reg3 after 5 ns when "011",
                reg3 after 5 ns when others;
with addr2 select
reg_out_2  <= reg0 after 5 ns when "000",
                reg1 after 5 ns when "001",
                reg2 after 5 ns when "010",
                reg3 after 5 ns when "011",
                reg3 after 5 ns when others;
end behavior;
```

FIGURE 3-6 Selected signal assignment statement

the action to be taken if addr1 or addr2 takes on any of the eight values including those between 4 and 7. This is not really restrictive, since in practice we must consider what would happen in the physical system in this case. The **others** keyword is used to state the value of the target signal over a range of values and thereby cover the whole range. Finally, the select expression can be quite flexible, for example, incorporating Boolean expressions.

As with simple and conditional CSAs, we must be aware of the conditions under which a selected signal assignment statement is executed. When an event occurs on a signal used in the select expression or any of the signals used in one of the choices, the statement is executed. This follows the expected behavior of the corresponding physical implementation where an event on any of the addresses or register contents could potentially change the value of the output signal.

Note a few new statements in this example. First, we initialize the values of the registers when they are declared. In this example, the even-numbered registers are initialized with a hexadecimal value denoted by x"12345678" while the odd numbered registers are initialized with the hexadecimal value denoted by x"abcdef00". Note that the target is a signal of type `std_logic_vector`. The hexadecimal values must be converted to the type `std_logic_vector` before they can be assigned. On the other hand, if the values were specified in binary notation explicit type conversion would not be required. The function `to_stdlogicvector` () is in the package `std_logic_1164` and performs this type conversion operation. While this was necessary of the simulator we used (Workview Office), this is not necessarily the case for other simulators (e.g., Cypress WARP2). To promote portability in this text we retain explicit type conversion where necessary in all of the examples. Type conversion as well as other functions in support of the IEEE 1164 value system are found in the `std_logic_1164` package provided by practically all CAD tool vendors. There are also other packages of functions and procedures that are provided by the vendors. Many standardization efforts are underway within the community in an effort to ensure portability of models between vendor toolsets and cooperating designers. These packages also provide similar type conversion functions often with slightly different names. Check availability of such packages within your toolset and browse through them. Packages may be located in the library `IEEE`. But be aware that in some systems they may be located in another design library. In this case a new **library** clause is required to declare this library, and the **use** clause must be appropriately modified to reference this library rather than `IEEE`! The use of libraries and packages is presented in greater detail in Chapter 6.

3.4   Constructing VHDL Models Using CSAs

Armed with concurrent signal assignment statements, we are now ready to construct VHDL models of interesting classes of digital systems. This section provides a prescription for constructing such VHDL models. By following this mechanical approach to constructing models, we can generate an intuition about the structure of VHDL programs and the utility of the language constructs discussed so far.

In a VHDL model written using only concurrent signal assignment statements, the execution of a signal assignment statement is initiated by the flow of data or signal values rather than the textual order of the statements. Based on the language features we have seen thus far, a model of a digital system will be comprised of an entity–architecture pair. The architecture model, in turn, will be comprised of some combination of simple, conditional, and selected signal assignments statements. The architecture may also declare and use internal signals in addition to the input and output ports declared in the entity description.

The following description assumes we are writing a VHDL model of a gate level, combinational circuit. However, the approach can certainly be applied to higher-level systems using combinational building blocks such as encoders and multiplexors. The simple

Input ports   Output ports

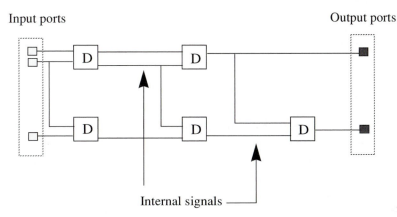

Internal signals

FIGURE 3-7 Delay element model of a digital system

methodology comprises two steps: (i) the drawing of an annotated schematic, and (ii) the conversion to a VHDL description. The following procedure outlines a few simple steps to organize the information we have about the physical system prior to writing the VHDL model.

Construct_Schematic

1. Represent each component (e.g., gate) of the system to be modeled as a *delay element*. The delay element simply captures all of the delays associated with the computation represented by the component and propagation of signals through the component. For each output signal of a component associate a specific value of delay through the component for that output signal.

2. Draw a schematic interconnecting all of the delay elements. Uniquely label each component.

3. Identify the input signals of the circuit as input ports.

4. Identify the output signals of the circuit as output ports.

5. All remaining signals are internal signals.

6. Associate a type with each input, output, and internal signal, such as `std_logic` or `std_logic_vector.`

7. Ensure that each input port, output port, and internal signal are labeled with a unique name.

An example of such a schematic is shown in Figure 3-7. Now, from this schematic, we can write a VHDL model using concurrent signal assignment statements. A template for the VHDL description is shown in Figure 3-8 This template can be filled in as described below in the procedure **Construct_CSA_Model**.

Construct_CSA_Model

1. At this point I recommend using the IEEE 1164 value system. To do so, include the following two lines at the top of your model declaration.

   library IEEE;
   use IEEE.std_logic_1164.all;

   Single-bit signals can be declared to be of type std_logic while multibit quantities can be declared to be of type std_logic_vector.

2. Select a name for the entity (entity_name) and write the entity description specifying each input or output signal port, its mode, and associated type. This can be read off of the annotated schematic.

3. Select a name for the architecture (arch_name) and write the architecture description. Place both the entity and architecture descriptions in the same file (as we will see in Chapter 8 this is not necessary.)

   3.1 Within the architecture description, name and declare all of the internal signals used to connect the components. The declaration states the type of each signal and its initial value. Initialization is not required, but is recommended. These declarations occur prior to the first **begin** statement in the architecture.

   3.2 Each internal signal is driven by exactly one component. If this is not the case make sure the type of the signal is a resolved type such as std_logic or std_logic_vector. For each internal signal, write a concurrent signal assignment statement that expresses the value of this internal signal as a function of the component input signals that are used to compute its value. Use the delay value associated with that output signal for that component.

   3.3 Each output port signal is driven by the output of some internal component. For each output port signal write a concurrent signal assignment statement that expresses its value as some function of the signals that are inputs to that component.

   3.4 If you are using any functions or type definitions provided by a third party make sure that you have declared the appropriate library using the **library** clause and declared the use of this package via the presence of a **use** clause in your model.

   If there are S signals and ports in the schematic, there will be S concurrent signal assignment statements in the VHDL model—one for each signal. This approach provides a quick way of constructing VHDL models by attempting to maintain a close correspondence with the hardware being modeled. There are many alternatives to constructing a VHDL model and the above approach represents only one method. With experience, the reader will no doubt discover many other alternatives for constructing efficient models for digital systems of interest.

library library-name-1, library-name-2;

use library-name-1.package-name.all;

use library-name-2.package-name.all;

entity entity_name **is**

port(*input signals* : **in** *type*;

    output signals : **out** *type*);

end entity_name;

architecture arch_name **of** entity_name **is**

-- declare internal signals

-- you may have multiple signals of different types

signal internal-signal-1 : *type* := *initialization*;

signal internal-signal-2 : *type* := *initialization*;

begin

-- specify value of each signal as a function of other signals

internal-signal-1 <= *simple, conditional, or selected CSA*;

internal-signal-2 <= *simple, conditional, or selected CSA*;

output-signal-1 <= *simple, conditional, or selected CSA*;

output-signal-2 <= *simple, conditional, or selected CSA*;

end arch_name;

FIGURE 3-8 A template for writing VHDL models using CSAs

Simulation Exercise 3.2: A One-Bit ALU

Consider a simple one-bit ALU shown in Figure 3-9 that performs the AND, OR, and ADDITION operations. The result produced at the ALU output depends on the value of signal OPCODE. Write and simulate a model of this ALU using concurrent signal assignments statements. Test each OPCODE to ensure that the model is accurate by examining the waveforms on the input and output signals. Use a gate delay of 5 ns and a delay of 2 ns through the multiplexor. Remember, while the OPCODE field is two bits wide, there are only three valid inputs to the multiplexor.

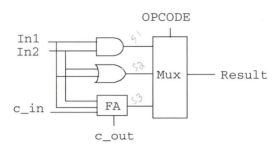

FIGURE 3-9 A single-bit ALU (FA = full adder)

Step 1. Follow the steps in **Construct_Schematic**. Ensure that all of the signals including the input and output ports are defined, labeled, and their mode and types are specified.

Step 2. Follow the steps in **Construct_CSA_Model**. To describe the operation of the full adder, use two simple concurrent signal assignment statements: one each to describe the computation of the `sum` and `carry` outputs respectively. Call this file *alu.vhd*

Step 3. Compile *alu.vhd*.

Step 4. Load the simulation model into the simulator.

Step 5. Generate a sequence of inputs that you can use to verify that the model is functioning correctly.

Step 6. Open a trace window with the signals you would like to trace. Include internal signals which are signals that are not entity ports in the model.

Step 7. Run the simulation for 50 ns.

Step 8. Check the trace to determine correctness.

Step 9. Print and record the trace.

Step 10. Add new operations to the single-bit ALU, recompile, and resimulate the model. For example, you can add the exclusive-OR, subtraction, and complement operations.

End Simulation Exercise 3.2

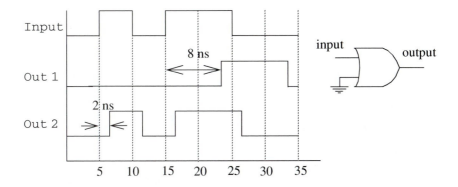

FIGURE 3-10 An example of the inertial delay model. Out 1 is the output waveform for delay = 8 ns and Out 2 is the output waveform for delay = 2ns.

3.5 Understanding Delays

We now have a template to help us begin to write basic VHDL models for many digital circuits. Let us examine one important aspect of these models in greater detail: propagation delays. Accurate representation of the behavior of digital circuits requires accurate modeling of delays through the various components. This section discusses the delay models available in VHDL and how they are specified. These models can be incorporated in a straightforward manner into the basic template for writing VHDL models that was described earlier. They are also used in the models that are described in Chapter 4 and the structural models that are described in Chapter 5.

3.5.1 The Inertial Delay Model

In Chapter 2 it was pointed out that digital circuits have a certain amount of inertia. For example, it takes a finite amount of time and a certain amount of energy for the output of a gate to respond to a change on the input. This implies that the change on the input has to persist for a certain period of time to ensure that the output will respond. If it does not persist long enough the input events will not be propagated to the output. This propagation delay model is referred to as the *inertial delay model* and is the default delay model for VHDL programs.

An example is shown in Figure 3-10 where a signal is applied to the input of a 2-input OR gate. If the gate delay is 8 ns, any pulse on the input signal of duration less than 8 ns will not be propagated to the output. This is illustrated by waveform Out 1. However, if the gate delay is 2 ns we see that each pulse on the input waveform is of a duration greater than 2 ns and is therefore propagated to the output. Any pulse with a width of less than the propagation delay through the gate is said to be rejected. In general, the pulse widths that are actually rejected in a physical circuit are very dependent upon the physical design and manufacturing process parameters and can be difficult to determine accurately. The VHDL language uses the propagation delay through the component as the default pulse rejection width.

VHDL'93   However, if we have a greater understanding of the properties of the components that we are modeling, the VHDL'93 revision supports the following specification of a value for the pulse rejection width.

> sum <= **reject** 2 **ns inertial** (a **xor** b) **after 5 ns**;

In Section 3.3.2, waveform elements were introduced as the manner in which we could specify the new value–time pair of a signal. The definition of a waveform element is extended in VHDL'93 to allow us to specify a distinct pulse rejection width (distinct from the propagation delay). Note that the expression has been preceded by the keyword **reject**

VHDL'93 and a time value has been provided. The keyword **inertial** must also be provided. Thus, for VHDL'93 we can write the general form of the simple concurrent signal assignment statement as follows.

> *signal* <= **reject** *time-expression* **inertial** *value-expression* **after** *time-expression*;

The statement is a general form describing the occurrence of an event on a signal. This form specifies the value of the signal, the time at which the signal is to receive this value, and the duration over which the input pulse must persist if the output is to receive this new value.

3.5.2 The Transport Delay Model

Like switching devices, signals propagate through wires at a finite rate and experience delays that are proportional to the distance. However, unlike switching devices, wires have comparatively less inertia. As a result, wires will propagate signals with very small pulse widths and we can model wires as media that will propagate any changes in signal values independent of the duration of the pulse width. In modern technologies with increasingly small feature sizes the wire delays dominate, and designs seek to minimize wire length. In these circuits wire delays are non-negligible and should be modeled to produce accurate simulations of circuit behavior. Such delays are referred to as *transport delays*. As with inertial delays, these delays can be specified by prefacing a waveform element with the keyword **transport** as follows:

> sum <= **transport** (a **xor** b) **after 5 ns**;

In this case a pulse of any width on signal a or b will be propagated to the sum signal. We will generally not use the transport delay model for modeling components that have significant inertia.

Example: Transport Delays

Up to this point, digital components have been treated as delay elements. Output signals acquire values after a specified propagation delay that we now know can be specified to be an inertial delay or a transport delay. If we wish to model delays along wires we can simply replace the wire with a delay element. The delay value is equal to the delay experienced by the signal transmission along the wire and the delay type is transport. Consider the half-adder circuit again redrawn to capture wire delays on the output signals, as shown in Figure 3-11. Delay elements model the delay on the sum and carry signals. Note in this example that the delay along these wires is longer than the propagation delay through the gate. From the timing diagram we see that a pulse of width 2 ns on the sum input at time 0 is propagated to signal s1 at time 2 ns. The wire delay is 4 ns. Under the inertial delay model this pulse would be rejected and would not appear on the sum output. However, we have specified the delay type to be transport. Therefore, the pulse is transmitted to the sum output after a delay of 4 ns. This signal is now delivered to the next circuit, having been delayed by an amount equal to the propagation delay through the signal wires. This approach enables the accurate modeling of wire delays although in practice it is difficult to obtain accurate estimates the wire delay without proceeding through physical design and the layout of the circuit.

Example End: Transport Delays

The choice between the use of inertial delay or transport delay is determined by the components that are being modeled. For example, if we have a model of a board level design we may have VHDL models of the individual chips. The delay experienced by signals between chips can be modeled using the transport delay model. The procedure **Construct_Schematic** provided in Section 3.4 can be modified to include delay elements for all wire delays to be modeled. The procedure for translating this annotated schematic is simply modified to use transport delays in the concurrent signal assignment statements that represent wire delays.

3.5.3 Delta Delays

What happens if we do not specify a delay for the occurrence of an event on a signal? For example, the computation of the outputs for an exclusive-OR gate may be written as follows:

```
sum <= (a xor b);
```

We may chose to ignore delays when we do not know what they are or when we are interested only in creating a simulation that is functionally correct and is not concerned with the timing behavior. For example, consider the timing of the full-adder model shown in Figure 3-4. There is a correct ordering of events on the signals. Input events on signals In1, In2, and c_in produce events on internal signals s1, s2, and s3, which, in turn,

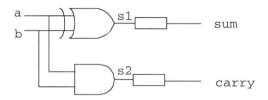

```
library IEEE;
use IEEE.std_logic_1164.all;
entity half_adder is
port(a, b: in std_logic;
     sum, carry: out std_logic);
end half_adder;

architecture transport_delay of half_adder is
signal s1, s2: std_logic:= '0';
begin
s1 <= (a xor b) after 2 ns;
s2 <= (a and b) after 2 ns;
sum <= transport  s1 after 4 ns;
carry <= transport  s2 after 4 ns;
end transport_delay;
```

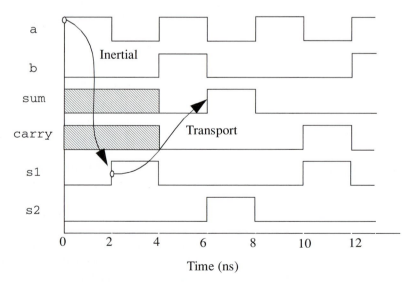

FIGURE 3-11 An example using transport delays

produce events on the output signals `sum` and `c_out`. For functional correctness we must maintain this ordering even when delays remain unspecified. This is achieved within the VHDL language by defining an infinitesimally small delay referred to as a *delta delay*. The above form of the signal assignment statement implicitly places a "**after** 0 **ns**" time expression following the value expression. When this is the case, the component is effectively assigned a delay value of Δ. Now simulation proceeds exactly as described in the earlier examples. As the following example will demonstrate, Δ does not actually have to be assigned a value but is utilized within the simulator to order events. If events with zero delay are produced at timestep T, the simulator simply organizes and processes events in time order of occurrence: events that occur Δ seconds later are followed by events occurring 2Δ seconds later, followed by events occurring 3Δ seconds later, and so on. Delta delays are simply used to enforce dependencies between events and thereby ensure correct simulation. The following example will help clarify the use of delta delays.

Example: Delta Delays

Consider the combinational logic circuit and the corresponding VHDL code shown in Figure 3-12. The model captures a behavioral description of the circuit *without* specifying any gate delays. Figure 3-13(a) illustrates the timing of the circuit when inputs are applied as shown. At time 10 ns the signal `In2` makes a $1 \rightarrow 0$ transition. This causes a sequence of events in the circuit resulting in the value of Z = 0. From the accompanying VHDL code we see that the gate delays are implicitly 0 ns. Therefore, the timing diagram shows the signal Z acquiring this value at the same instant in time that `In2` makes a transition. From the timing diagram it is also clear that signals s2 and s3 also make transitions at this instant in time. In reality, from the circuit diagram we know that there is a dependency between `In2` and s3, and s3 and Z. These dependencies are evident from the structure of the circuit. The event on `In2` precedes and causes the transitions on s3 and s4, while the transition on s3 causes the transition on Z. The simulation of the circuit honors these dependencies through the logical use of delta delays. The dependencies between `In2`, s3, and Z are shown in Figure 3-13(b). The transition on `In2` causes s3 to make a transition to 1 after Δ secs. The event on s3 causes a $1 \rightarrow 0$ event on Z after 2Δ secs. These events are referred to as delta events and are shown in Figure 3-13(b). These delta events take place within the simulator and do not appear on the external trace produced for the viewer (i.e., Figure 3-13(a)). The actual implementation of delta events is managed within the simulator by keeping track of signal values and when they are updated.

By forcing all events to take some infinitesimally small amount of time the dependencies between events can be preserved and the correct operation of the circuit maintained. Recall how the circuit is simulated. Time is advanced to that of the first event on

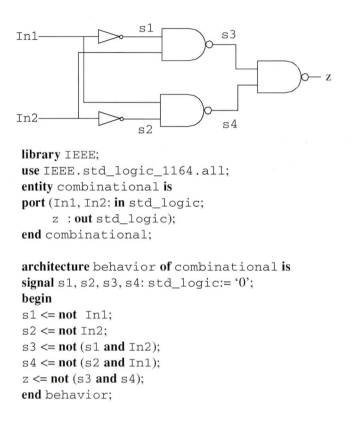

```
library IEEE;
use IEEE.std_logic_1164.all;
entity combinational is
port (In1, In2: in std_logic;
    z  : out std_logic);
end combinational;

architecture behavior of combinational is
signal s1, s2, s3, s4: std_logic:= '0';
begin
s1 <= not  In1;
s2 <= not In2;
s3 <= not (s1 and In2);
s4 <= not (s2 and In1);
z <= not (s3 and s4);
end behavior;
```

FIGURE 3-12 A VHDL model with delta delays

the list. The signal is assigned this value, any new outputs computed, and the process repeated. When time is advanced by Δ, this step is referred to as a delta cycle.

Example End: Delta Delays

Simulation Exercise 3.3: Delta Delays

Repeat the simulation of the full-adder model in Simulation Exercise 3.1, but do not specify any gate delays.

Step 1.   Run the simulation for 40 ns and trace input, internal, and output signals.

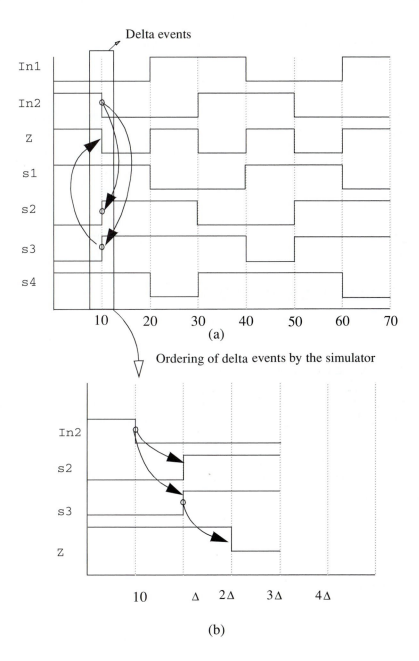

FIGURE 3-13 (a) Occurrence of delta events (b) Ordering imposed on these events within the simulator

Step 2. Annotate the trace to identify delta events.

Step 3. Compare the trace generated here with that generated in Simulation Exercise 3.1. What are the differences?

Step 4. Modify the model to include a 2 ns wire delay for internal signals. Recompile the model.

Step 5. Simulate the model now and generate another trace. The effect of wire delays is now explicitly captured.

Step 6. Identify events that occur in this second trace that are different from those in the earlier trace.

Step 7. Create an input stimulus for one of the inputs with pulses whose duration are both shorter and longer than the gate delay. Set the value of the remaining inputs to logic 0. This is usually achieved by adjusting the simulator step time and via stimulus commands unique to the simulator that you are using.

Step 8. Generate a trace and identify pulses that are rejected by the gate models.

End Simulation Exercise 3.3

3.6 Chapter Summary

The reader should be comfortable with the following concepts that have been introduced in this chapter. A syntactic reference to common language types and operators can be found in Chapter 9.

- Entity and architecture constructs
- Concurrent signal assignment statements
    - simple concurrent signal assignment
    - conditional concurrent signal assignment
    - selected concurrent signal assignment
- Constructing models using concurrent signal assignment statements
    - modeling events, propagation delays, and concurrency
- Modeling delays
    - inertial delay
    - transport delay
    - delta delay
- Signal drivers and projected waveforms
- Shared signals, resolved types, and resolution functions

- Generating waveforms using waveform elements
- Events and transactions

The VHDL models of systems should now be beginning to take some form. The reader should be capable of constructing functionally correct models for many types of digital systems utilizing inertial, transport, or delta delay models (i.e., functional models).

Exercises

1. A good exercise for understanding entity descriptions is to write the entity descriptions for components found in data books from component vendors, for example, the TTL data book. These entity descriptions can be compiled without the architecture descriptions and thus can be checked for syntactic correctness. Of course, we cannot say anything about the semantic correctness of such descriptions since we have not even written the architecture descriptions yet!

2. Write and simulate a VHDL model of a 2-bit comparator.

3. Sketch the output waveform produced by the following VHDL simple concurrent signal assignment statements.

   s1 <= '0' **after** 5 **ns**, '1' **after** 15 **ns**, '0' **after** 35 **ns**, '1' **after** 50 **ns**;

   s1 <= '0' **after** 20 **ns**, '1' **after** 25 **ns**, '0' **after** 50 **ns**;

4. Construct and test VHDL modules for generating the periodic waveforms with the structure shown in Figure 3-14.

5. Construct a VHDL model that will accept a clock signal as input and produce the complement signal as output with a delay of 10 ns.

6. Write and simulate the entity–architecture description of a 3-bit decoder using the conditional signal assignment statement. Test the model with all possible combinations of inputs and plot the decoder output waveform.

7. Repeat the preceding exercise by building the decoder from basic gates. Is there any difference in the number of events generated between the simulation of this model and a model where the behavior is described using the conditional signal assignment statement? You should be able to answer this question by examining the traces in both cases over the same time interval. If your simulator permits, examine the event queues during simulation.

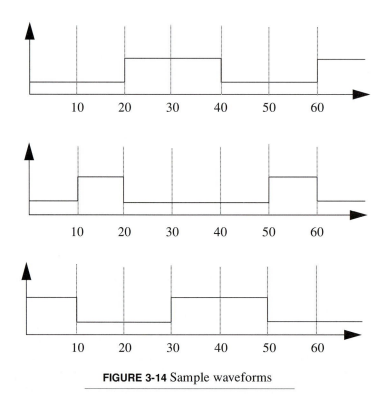

FIGURE 3-14 Sample waveforms

8. Write a VHDL model of the circuit shown below including wire delays, Use the transport delay model for the wire delays and assume a wire delay of 2 ns between components. Generate a timing diagram. Select your own gate delays. Mark events that would not have occurred under the default inertial delay model.

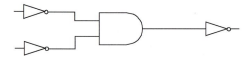

9. Why are the concepts of delta events and delta delays necessary for the correct discrete event simulation of digital circuits?

CHAPTER 4   Modeling Behavior

This chapter expands upon the approach described so far that uses concurrent signal assignment statements for constructing VHDL models. In Chapter 3 components were modeled as delay elements and their internal behavior was described using concurrent signal assignment statements. In this chapter, we discuss more powerful constructs for describing the internal behavior of components when they cannot be simply modeled as delay elements. The basis for these descriptions is the *process* construct that enables us to use conventional programming language constructs and idioms. As a result we can model the behavior of components much more complex than delay elements and are able to model systems at higher levels of abstraction.

4.1 The Process Construct

The VHDL language and modeling concepts described in the previous chapter were derived from the operational characteristics of digital circuits, where the design is represented as a schematic of concurrently operating components. Each component is characterized by the generation of events on output signals in response to events on input signals. These output events may occur after a component dependent propagation delay. The component behavior is expressed using a CSA statement that explicitly relates input signals, output signals, and propagation delays. Such models are convenient to construct when components correspond to gates or switch level models of transistors. However, when we wish to construct models of complex components such as CPUs, memory modules, or communication protocols, such a model of behavior can be quite limiting. The event model is still valid—externally we see that events on the input signals will eventually cause events on the output signals of the component. However, the computation of the time at which these output events will occur and the value of the output signals can be quite complex. Moreover, for modeling components such as memories, we need to retain state information within the component description. It is not sufficient to be able to compute the values of the output signals as a function of the values of the input signals.

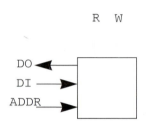

R   W

DO
DI
ADDR

FIGURE 4-1 A model of memory

For example, consider the behavior of a simple model of a memory module as shown in Figure 4-1. The memory module is provided with address, data, and read and write control signals. Let us assume that it contains 4096, 32-bit words of memory. The value of the R or W control signals determine whether the data on DI is to be written at the address on the ADDR port, or whether data is to be read from that address and provided on the output port DO. The signal S is used to signal completion of the memory operation. Events on the input address, data, or control lines produce events that cause the memory model to be executed. We can also reasonably expect to know the memory access times for read and write operations and therefore know the propagation delays. However, the behavior internal to the memory module is difficult to describe using only the signal assignment statements provided in Chapter 3. How can we represent memory words? How can we address the correct word based on the values of the address lines and control signals? The answers are easier to come by if we have access to conventional sequential programming language constructs. Memory can be implemented as an array and the address value can be used to index this array. Depending upon the value of the control signals we can decide if this array element is to be written or read. Such behavior can be realized in VHDL by using *sequential statements* via the *process* construct.

In contrast to concurrent signal assignment statements a process is a sequentially executed block of code. A VHDL model of a memory module equivalent to the one in Figure 4-1 is shown in Figure 4-2. This model consists of one process that is labeled mem_process. Process labels are delimited by colons. The structure of a process is very similar to that of programs written in a conventional block structured programming language such as Pascal. The process begins with a declarative region followed by the process body delimited by **begin** and **end** keywords. Variables and constants used within the process are declared within the declarative region. The **begin** keyword denotes the start of the computational part of the process. All of the statements in this process are executed sequentially. Data structures may include arrays and queues, and programs may use standard data types such as integers, characters, and real numbers. Unlike signals whose changes in values must be scheduled to occur at discrete points in time, variable assignments take effect immediately. Variable assignment is denoted by the ":=" operator. Since all statements are executed sequentially within a process, values assigned to variables are

```vhdl
library IEEE;
use IEEE.std_logic_1164.all;
use WORK.std_logic_arith.all;
                    -- we need this package for 1164 related functions
entity memory is
port (address, write_data : in std_logic_vector (31 downto 0);
    MemWrite, MemRead : in std_logic;
    read_data  : out std_logic_vector (31 downto 0));
end memory;

architecture behavioral of memory is
type mem_array is array(0 to 7) of std_logic_vector (31 downto 0);

begin
mem_process: process (address, write_data)
variable data_mem : mem_array := (
to_stdlogicvector(X"00000000"),   --- initialize data memory
to_stdlogicvector(X"00000000"),   ---
to_stdlogicvector(X"00000000"),   ---
to_stdlogicvector(X"00000000"),
to_stdlogicvector(X"00000000"),
to_stdlogicvector(X"00000000"),
to_stdlogicvector(X"00000000"),
to_stdlogicvector(X"00000000"));
variable addr :integer;

begin

-- the following type conversion function is in std_logic_arith

addr := to_integer (address  (2 downto 0));
if MemWrite = '1' then
data_mem(addr) := write_data;

elsif MemRead  = '1' then
read_data <= data_mem(addr) after 10 ns;
end if;
end process mem_process;
end behavioral;
```

FIGURE 4-2 A behavioral description of a memory module

visible to all following statements within that same process. Control flow within a process is strictly sequential altered by constructs such as **if-then-else**, or **loop** statements. In fact, we can regard the process itself as a traditional sequential program. One of the distinguishing features of processes is that we can make assignments to signals declared external to the process. For example, consider the memory model in Figure 4-2. At the end of this process we have signal assignment statements that assign internally computed values to signals in the interface after a specified propagation delay. Thus, externally we are able to maintain the discrete event execution model: events on the memory inputs produce events on the memory outputs after a delay equal to the memory access time. However, internally we are able to develop complex models of behavior that produce these external events. *With respect to simulation time a process executes in zero time.* Delays are associated only with the assignment of values to signals.

Recall that a CSA is executed any time an event occurs on a signal in the right-hand side of the signal assignment statement. When is a process executed? In Figure 4-2, adjacent to the **process** keyword is a list of input signals to the component. This list is referred to as the *sensitivity list.* The execution of a process is initiated whenever an event occurs on any of the signals in the sensitivity list of the process. Once started, the process executes to completion in zero (simulation) time and potentially generates a new set of events on output signals. We begin to see the similarity between a process and a CSA. In the models with concurrent signal assignment statements, input signals are inferred by their presence on the right-hand side of the signal assignment statement. In a process the signals can be listed in the sensitivity list. For all practical purposes, we can regard a process simply as a "big" concurrent signal assignment statement that executes concurrently with other processes and signal assignment statements. Processes are simply capable of describing more complex events than the CSAs described in Chapter 3. *In fact, CSAs are themselves processes!* However, they are special and do not require the **process, begin,** and **end** syntax of more complex processes.

You will notice the use of another package in the model shown in Figure 4-2: `std_logic_arith`. The definition of the type conversion function `to_integer()` is in this package. This function is necessary for the following reason. Memory is modeled as an array of 32-bit words. This array is indexed by an integer. Therefore, the memory address that is provided as a 32-bit number of type `std_logic_vector` must be converted to an integer before this array can be accessed. Of course we do not create an array with 2^{32} entries but rather eight words of memory. Therefore, the model uses only the lower 3 bits of the memory address. In this model and a few others in this text we have used the package `std_logic_arith`. We have compiled this model into the library `WORK`. This library is the default working directory and is implicitly declared. We do not need a **library** clause in our model. Many vendors will support various packages with many useful type conversion, arithmetic, and logic functions. These packages will be placed in various libraries. Check with your installation to determine the location and contents of available packages. This is all we need to know for the moment. We will revisit packages and libraries in greater detail in Chapter 6.

Since statements within a process are executed sequentially these statements are referred to as *sequential statements,* in contrast to the concurrent signal assignment statements that we saw in Chapter 3. Processes can be thought of as programs that are executed within the simulation to model the behavior of a component. Thus, we have more powerful means to model the behavior of digital systems. Such models are often referred to as behavioral models, although any VHDL model using concurrent or sequential statements is a description of behavior.

Once the concepts of a process and the underlying semantics are understood, we need to know the syntax of the major programming constructs that we can use within a process. Identifiers, operators, and useful data types are provided in Chapter 9. Based on our experience with other high level-languages, we can begin immediately describing the behavior of components and developing nontrivial simulation models.

4.2 Programming Constructs

4.2.1 If-Then-Else

An **if** statement is executed by evaluating an expression and conditionally executing a block of sequential statements. The structure may optionally include an **else** component. The statement may also include zero or more **elsif** branches (note the absence of the letter 'e' in **elsif**). In this case all of the Boolean valued expressions are evaluated sequentially until the first true expression is encountered. An **if** statement is closed by the **end if** clause. A good example of the utility of the use of the **if-then-else** construct is captured in the memory model of Figure 4-2.

4.2.2 Concurrent Processes and the Case Statement

The behavioral model shown in Figure 4-2 utilizes a single process. Just as we had concurrent signal assignment statements we may also have concurrently executing processes. Consider another behavioral model of a half adder with two processes as shown in Figure 4-3. Both processes are sensitive to events on the input signals a and b. Whenever an event occurs on either a or b, both processes are activated and execute concurrently in simulation time. The second process is structured using a **case** statement. The case statement is used whenever it is necessary to select one of several branches of execution based on the value of an expression. The branches of the **case** statement must cover all possible values of the expression being tested. Each value of the case expression being tested can belong to only one branch of the case statement. The **others** clause can be used to ensure that all possible values for the case expression are covered. Although this example shows a single statement within each branch, in general the branch can be comprised of a sequence of sequential statements. This example also shows that port signals are visible within a process. This means that process statements can read port values and schedule values on output ports.

```
library IEEE;
use IEEE.std_logic_1164.all;

entity half_adder is
port (a, b : in std_logic;
      sum, carry : out std_logic);
end half_adder;

architecture behavior of half_adder is
begin
sum_proc: process(a,b)
      begin
        if (a = b) then
            sum <= '0' after 5 ns;
        else
            sum <= (a or b) after 5 ns;
        end if;
      end process;

carry_proc: process (a,b)
      begin
        case a is
        when '0' =>
        carry <= a after 5 ns;
        when '1' =>
        carry <= b after 5 ns;
        when others =>
        carry <= 'X' after 5 ns;
        end case;
end process carry_proc;
end behavior;
```

FIGURE 4-3 A two-process half-adder model

4.2.3 Loops

There are two forms of the loop statement. The first form is the use of **for** loops. An example of the use of such a loop construct is shown in Figure 4-4. This example multiplies two 32-bit numbers by successively shifting the multiplicand and adding to the partial product if the corresponding bit of the multiplier is 1 [9]. The model simply implements what we have traditionally known as long multiplication using base 2 arithmetic. The model saves storage by using the lower half of the 64-bit product register to initially store the multiplier. As successive bits of the multiplier are examined, the bits in the lower half of the product register are shifted out, eventually leaving a 64-bit product. Note the use of the **&** operator representing concatenation. A logical shift right operation is specified by copying the upper 63 (out of 64) bits into the lower 63 bits of the product register and setting the most significant bit to 0 using the concatenation operator.

There are several unique features of this form of a loop statement. Note that the loop index is not declared anywhere within the process! The loop index is automatically and implicitly declared by virtue of its use within the loop statement. Moreover, the loop index is declared locally for this loop. If a variable or signal with the name `index` is used elsewhere within the same process or architecture (but not in the same loop), it is treated as a distinct object. Unlike other languages the loop index cannot be assigned a value or altered in the body of the loop. Therefore, loop indices cannot be provided as parameters via a procedure call or as an input port. We see that the loop index is exactly that, a loop index, and the language prevents us from using it in any other fashion. This does make it convenient to write loops, since we do not have to worry about our choice of variable names for the loop index conflicting with those elsewhere in the model.

Often it is necessary to continue the iteration until some condition is satisfied rather than performing a fixed number of iterations. A second form of the loop statement is the use of the **while** construct. In this form, the **for** statement is simply replaced by

> **while** (*condition*) **loop**

Unlike the **for** construct, the condition may involve variables that are modified within the loop. For example the **for** loop in Figure 4-4 could be replaced by the following.

> **while** j < 32 **loop**
> ...
> ...
> j := j+1;
> **end loop**;

4.3 More on Processes

Upon initialization all processes are executed once. Thereafter, processes are executed in a data-driven manner: activated by events on signals in the sensitivity list of the process or by waiting for the occurrence of specific events using the **wait** statement (described in Section 4.4). Remember the sensitivity list of a process is not a parameter list! This list simply identifies those signals to which the process is sensitive: when an event occurs on any one of these signals, the process is executed. This is analogous to CSAs, which are executed whenever an event occurs on a signal on the right-hand side of a CSA. In fact CSAs are really processes with simpler syntax. All of the ports of the entity and the signals declared within an architecture are visible within a process, which means that they can be

```vhdl
library IEEE;
use IEEE.std_logic_1164.all;
use WORK.std_logic_arith.all; -- needed for arithmetic functions

entity mult32 is
port (multiplicand, multiplier : in std_logic_vector(31 downto 0);
     product : out std_logic_vector (63 downto 0));
end mult32;

architecture behavioral of mult32 is
constant module_delay: Time:= 10 ns;
begin
mult_process: process(multiplicand, multiplier)
variable product_register : std_logic_vector (63 downto 0) :=
to_stdlogicvector(X"0000000000000000");
variable multiplicand_register : std_logic_vector(31 downto 0):=
to_stdlogicvector(X"00000000");

begin
multiplicand_register := multiplicand;
product_register(63 downto 0) :=
to_stdlogicvector(X"00000000") & multiplier;
--
-- repeated shift-and-add loop
--
for index in 1 to 32 loop
if product_register(0) = '1' then
product_register(63 downto 32) := product_register(63 downto 32) +
multiplicand_register(31 downto 0);
end if;
            -- perform a right shift with zero fill
product_register  (63 downto  0) := '0' & product_register (63
downto 1);
end loop;
-- write result to output port
product <= product_register after module_delay;

end process mult_process;
end behavioral;
```

FIGURE 4-4 An example of the use of the loop construct

read or assigned values from within a process. Thus, during the course of execution a process may read or write any of the signals declared in the architecture or any of the ports on the entity. For example, this is how processes can communicate among themselves. Process A may write a signal that is in the sensitivity list of Process B. This will cause Process B to execute. Process B may in turn similarly write a signal in the sensitivity list of Process A. The use of communicating processes is elaborated in the following example.

Example: Communicating Processes

This example illustrates a model of a full adder constructed from two half adders and a two-input OR gate. The behaviors of the three components are described using processes that communicate through signals and is shown in Figure 4-5. When there is an event on either input signal, process HA1 executes, creating events on internal signals s1 and s2. These signals are in the sensitivity lists of processes HA2 and O1 and therefore these processes will execute and schedule events on their outputs as necessary. Note that this style of modeling still follows the structural description of the hardware where we have one process for each hardware component of Figure 4-5. Contrast this model with the model described in Figure 3-3.

Example End: Communicating Processes

Simulation Exercise 4.1: Combinational Shift Logic

This exercise is concerned with the construction of a combinational logic shifter. The inputs to the shift logic include a 3-bit operand specifying the shift amount, two single-bit signals identifying the direction of the shift operation—left or right—and an 8-bit operand. The output of the shift logic is the shifted 8-bit operand. Shift operations provide zero fill. For example, a left shift of the number 01101111 by 3 bit positions will produce the output 01111000.

Step 1. Create a text file with the entity description and the architecture description of the shift logic. Assume the delay through the shift logic is fixed at 40 ns, independent of the number of digits that are shifted. While you can implement this behavior in many ways, for this assignment use a single process and the sequential VHDL statements to implement the behavior of the shift logic. You might find it useful to use the concatenation operator, **&,** and addressing within arrays to perform the shift operations. For example, we can perform the following assignment

dataout <= datain(4 **downto** 0) & "000";

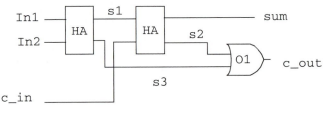

```
library IEEE;
use IEEE.std_logic_1164.all;

entity full_adder is
port (In1, c_in, In2 : in std_logic;
      sum, c_out : out std_logic);
end full_adder;

architecture behavioral of full_adder is
signal  s1, s2, s3 : std_logic;
constant delay :Time:= 5 ns;
begin
HA1: process (In1, In2) --process describing the first half adder
begin
s1 <= (In1 xor In2) after delay;
s3  <= (In1  and In2) after delay;
end process HA1;

HA2: process(s1,c_in) -- process describing the second half adder
begin
sum <= (s1  xor c_in) after delay;
s2 <= (s1 and c_in) after delay;
end process HA2;

OR1: process (s2, s3) -- process describing the two-input OR gate
begin
c_out <= (s2 or s3) after delay;
end process OR1;
end behavioral;
```

FIGURE 4-5 A communicating process model of a full adder

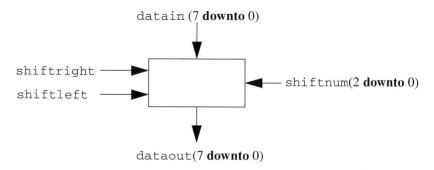

FIGURE 4-6 Interface description for a combinational logic shifter

This assignment statement will perform a left shift by three digits with zero fill. Both input and output operands are 8-bit numbers. For VHDL'93 you may use the VHDL built-in shift operators. Use the `case` statement to structure your process.

Step 2. Use the types `std_logic` and `std_logic_vector` for the input and output signals. Declare and reference the library `IEEE` and the package `std_logic_1164`.

Step 3. Create a sequence of test vectors. Each of the test vectors will specify the values of: (1) the `shiftright` and `shiftleft` single-bit control signals, (2) an 8-bit input operand, and (3) a 3-bit number that specifies the number of digits the input operand is to be shifted. Your test cases should be sufficient to ensure the model is operating correctly.

Step 4. Load the simulation model into the simulator. Set the simulator step time to be equal to the value of the propagation delay through the shift logic.

Step 5. Using the facilities available within the simulator, generate the input stimulus and open a trace window to view both the input stimulus and the output operand value.

Step 6. Exercise the simulator by running the simulation long enough to cover your test cases. Verify correct operation from the trace.

Step 7. Once you have the simulation functioning correctly, modify your model to implement circular shift operations. These operations are such that the digits that are shifted out of one end of the operand are the inputs to the other end. For example, a circular left shift of the pattern 10010111 by 3 digits will be 10111100. The circular shift operations can be implemented using the concatenation operator. In VHDL'93 the circular shift operations can be implemented with the VHDL predefined operators.

End Simulation Exercise 4.1

4.4 The Wait Statement

The execution behavior of the models presented in Chapter 3 and the behavioral models described so far in this chapter have been data-driven, where events on the input signals initiated the execution of processes. Processes will then suspend until the next event on a signal defined in its sensitivity list. This fits well with the behavior of combinational circuits, where a change on the input signals may cause a change in the value of the output signals. Therefore the outputs should be recomputed whenever there is a change in the value of the input signal.

However, what about modeling circuits where the outputs are computed only at specific points in time independent of events on the inputs? How do we model circuits which respond only to certain events on the input signals? For example, in synchronous sequential circuits, the clock signal determines when the outputs may change or when inputs are read. Such behavior requires us to be able to specify in a more general manner the conditions under which the circuit outputs must be recomputed. In VHDL terms, we need a more general way of specifying when a process is executed or suspended pending the occurrence of an event or events. This capability is provided by the **wait** statement.

The **wait** statements explicitly specify the conditions under which a process may resume execution after being suspended. The forms of the **wait** statement include the following.

> **wait for** *time expression*;

> **wait on** *signal*;

> **wait until** *condition*;

> **wait**;

The first form of the wait statement causes suspension of the process for a period of time given by the evaluation of *time expression*. This is an expression that should evaluate to a value that is of type **time**. The simplest form of this statement is as follows:

> **wait for** 20 **ns**;

The second form causes a process to suspend execution until an event occurs on one or more signals in a group of signals. For example, we might have the following statement.

> **wait on** `clk, reset, status`;

In this case, an event on any of the signals causes the process to resume execution with the first statement following the **wait** statement. The third form can specify a condition that evaluates to a Boolean value, TRUE or FALSE.

Using these wait statements processes can be used to model components that are not necessarily data driven but are driven only by certain types of events such as the rising edge of a clock signal. Many such conditions cannot be described using sensitivity lists alone. More importantly, we would often like to construct models where we need to suspend a process at multiple points within a process and not just at the beginning. Such models are made possible through the use of the **wait** statement. The following examples will help further motivate the use of the **wait** statement.

Example: Positive Edge-Triggered D Flip-Flop

The model of a positive edge-triggered D flip-flop is a good example of the use of the wait statement. The behavior of this component is such that the D input is sampled on the rising edge of the clock and transferred to the output. Therefore, the model description must be able to specify computations of output values only at specific points on time—in this case the rising edge of the clock signal. This is done using the **wait** statement, as shown in Figure 4-7. This brings us to another very interesting feature of the language. Note the statement clk'**event** in the model shown in Figure 4-7. This statement is true if an event (i.e., signal transition) has occurred on the clk signal. The conjunction, (clk'**event and** clk = '1'), is true for a rising edge on the clk signal. The signal clock is said to have an *attribute* named **event** associated with it. The predicate clk'**event** is true whenever an event has occurred on the signal clk in the most recent simulation cycle. Recall that an event is a change in the signal value. In contrast, a *transaction* occurs on a signal when a new assignment has been made to the signal, but the value may not have changed. As this example illustrates, such an attribute is very useful. The following section lists some useful attributes of VHDL objects.

The std_logic_1164 package also provides two useful functions that we could have used in lieu of attributes: rising_edge (clk) and falling_edge (clk). These functions take a signal of type std_logic as an argument and return a Boolean value denoting whether a rising edge (falling edge) occurred on the signal. The predicate clk'**event** simply denotes a change in value. Note that a single-bit signal of type std_logic can have up to nine values. Thus, if we are really looking for a rising edge from signal value 0 to 1, or a falling edge from signal value 1 to 0, it would be better to replace the test "**if** (clk'**event and** clk = '1')" with "**if** rising_edge(clk)."

Continuing with the description of the operation of the D flip-flop, we see that the input is sampled on the rising clock edge and the output values are scheduled after a period equal to the propagation delay through the flip-flop. The process is not executed whenever there is a change in the value of the input signal D, but rather only when there is a rising edge on the signal Clk.

Example End: Positive Edge-Triggered D Flip-Flop

The preceding example did not specify the initial values of the flip-flop. When a physical system is powered up the individual flip-flops may be initialized to some known state, but not necessarily all in the same state. In general, it is better to have some control over initial states of the flip-flops. This is usually achieved by providing such inputs as Clear or Set and Preset or Reset. Asserting the Set input forces Q = 1 and asserting the Reset input forces Q = 0. These signals override the effect of the clock signal and

```
library IEEE;
use IEEE.std_logic_1164.all;
entity dff is
port (D, Clk : in std_logic;
     Q, Qbar : out std_logic);
end dff;

architecture behavioral of dff is
begin
output: process
begin
wait until (Clk'event and Clk = '1');

    Q <= D after 5 ns;
    Qbar <= not D after 5 ns;

end process output;
end behavioral;
```

FIGURE 4-7 Behavioral model of a positive edge-triggered D flip-flop

are active at any time, hence the characterization as asynchronous inputs as opposed to the synchronous nature of the clock signal. The following example illustrates how we can extend the previous model to include asynchronous inputs.

Example: D Flip-Flop with Asynchronous Inputs

Figure 4-8 shows a model of a D flip-flop with asynchronous reset (R) and set (S) inputs and the corresponding VHDL model. The R input overrides the S input. Both signals are active low. Therefore, to set the output Q = 0, a zero pulse is applied to the reset input while the set input is held to 1, and vice versa. During synchronous operation, both S and R must be held to 1.

Example End: D Flip-Flop with Asynchronous Inputs

Now that we have seen how to create a model for a basic unit of storage with asynchronous inputs, it is relatively straightforward to similarly create models for registers and counters. Such an example is presented next.

```
library IEEE;
use IEEE.std_logic_1164.all;
entity asynch_dff is
port (R, S, D, Clk : in std_logic;
    Q, Qbar : out std_logic);
end asynch_dff;

architecture behavioral of asynch_dff is
begin
output: process  (R, S, Clk)
begin
if (R = '0')  then
    Q <= '0' after 5 ns;
    Qbar <= '1' after 5 ns;
elsif S = '0' then
    Q <= '1' after 5 ns;
    Qbar <= '0' after 5 ns;
  elsif (Clk'event and Clk = '1') then
    Q<= D after 5 ns;
    Qbar <= (not D) after 5 ns;
end if;
end process output;
end behavioral;
```

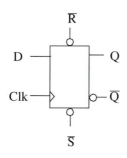

$\overline{S}$	$\overline{R}$	Clk	D	Q	$\overline{Q}$
0	1	X	X	1	0
1	0	X	X	0	1
1	1	R	1	1	0
1	1	R	0	0	1
0	0	X	X	?	?

FIGURE 4-8 D flip-flop with asynchronous set and reset inputs

```
library IEEE;
use IEEE.std_logic_1164.all;
entity reg4 is
port (D : in std_logic_vector (3 downto 0) ;
     Cl, enable, Clk: in std_logic;
     Q : out std_logic_vector (3 downto 0));
end reg4;

architecture behavioral of reg4 is
begin
reg_process: process (Cl, Clk)
begin
if (Cl = '1')  then
   Q <= "0000" after 5 ns;
   elsif (Clk'event and Clk = '1') then
   if enable = '1' then
   Q<= D after 5 ns;
   end if;
end if;
end process reg_process;
end behavioral;
```

FIGURE 4-9 A 4-bit register with asynchronous inputs and enable

Example: Registers and Counters

We can construct a model of a typical 4-bit register comprised of edge-triggered D flip-flops, with asynchronous clear and enable signals. Such a model is shown in Figure 4-9. With a few modifications this example can be converted into a model of a counter. Rather than sampling the inputs on each clock edge we can simply increment the value stored in the register. The initialization step can be also changed to load a preset value into the counter rather than initializing the counter to 0.

Example End: Registers and Counters

Example: Asynchronous Communication

Another example of the utility of wait statements is the modeling of asynchronous communication between two devices. A simple 4-phase protocol for synchronizing the transfer of data between a producer and consumer is shown in Figure 4-10. Let us assume the producer (e.g., input device such as a microphone) is providing data for a consumer device (e.g., the processor) and the transfer of each word must be synchronized. When the producer has data to be transferred the signal RQ is asserted. The consumer waits for a rising edge of the RQ signal before reading the data. The consumer then signals successful reception of the data by asserting the ACK signal. This causes the producer to de-assert RQ, which in turn results in the consumer de-asserting ACK. At this point the transaction is completed and the consumer and producer can each assert that the other has successfully completed their end of the transaction. The producer and consumer can be modeled as processes that communicate via signals. Such a model is shown in Figure 4-11. Note that signals RQ, ACK, and transmit_data are declared in the architecture and are visible within both processes even though they are not in the sensitivity list of either process. Although this example is incomplete in that the processes do not perform any interesting computations with the data, it does illustrate the use of wait statements to control asynchronous communication between processes. Moreover, it illustrates the ability to suspend the execution of a process at multiple points within the model. Since these processes execute concurrently in simulated time, they must be capable of suspending and resuming execution at multiple points within the VHDL code. Such a model is not possible using only sensitivity lists as the mechanism for initiating process execution (although other solutions to the producer–consumer problem are possible).

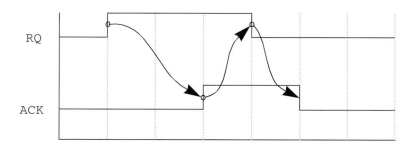

FIGURE 4-10 4-phase handshake

Example End: Asynchronous Communication

```vhdl
library IEEE;
use IEEE.std_logic_1164.all;

entity handshake is
port (input_data : in std_logic_vector(31 downto 0));
end handshake;

architecture behavioral of handshake is
signal transmit_data: std_logic_vector (31 downto 0);
signal RQ, ACK  : std_logic;
begin
producer: process
begin
wait until input_data'event;  --  wait until input data is available
transmit_data <= input_data;    -- provide data as producer
  RQ <= '1';
wait until ACK = '1';
RQ <= '0';
wait until ACK = '0';
end process producer;

consumer: process
variable receive_data : std_logic_vector (31 downto 0);
 begin
wait until RQ = '1';
receive_data := transmit_data; -- read data as consumer
ACK <= '1';
wait until RQ = '0';
ACK <= '0';
end process consumer;
end behavioral;
```

FIGURE 4-11 A VHDL model of the behavior shown in Figure 4-10

4.5 Attributes

The example of the model of the D flip-flop introduced the idea of an attribute of a signal. Attributes can be used to return various types of information about a signal. For example, in addition to determining if an event has occurred on the signal we might be interested in knowing the amount of time that has elapsed since the last event occurred on the signal. This attribute is denoted using the following syntax:

`clk`'**last_event**

In effect, when the simulator executes this statement a function call occurs that checks this property. The function returns the time since the last event occurred on signal `clk`. Such attributes are referred to as *function attributes*. Other useful signal attributes are shown in Table 4-1.

TABLE 4-1 Some useful function signal attributes

Function attribute	Function
`signal_name`'**event**	Function returning a Boolean value signifying a change in value on this signal
`signal_name`'**active**	Function returning a Boolean value signifying an assignment made to this signal. This assignment may not be a new value.
`signal_name`'**last_event**	Function returning the time since the last event on this signal
`signal_name`'**last_active**	Function returning the time since the signal was last active
`signal_name`'**last_value**	Function returning the previous value of this signal

There are several other useful classes of attributes, but only one other class will be described here: *value attributes*. As the name suggests, these attributes return values. Some commonly used value attributes are shown in Table 4-2.

For example, the memory model shown in Figure 4-2 contains the definition of a new type as follows.

type mem_array **is array**(0 **to** 7) **of** std_logic_vector (31 **downto** 0);

From Table 4-2 we have mem_array'**left** = 0, mem_array'**ascending** = true and mem_array'**length** = 8. Another useful example is in the use of enumerated types. For example, when writing models of state machines described later in this chapter, it is useful to have the following data type defined:

type statetype **is** (state0, state1, state2, state3);

In this case we have statetype'**left** = state0 and statetype'**right** = state3. This is useful in behavioral models when we wish to initialize signals to values based on their types. We may not always know the range and values of the various data types. The use of attributes makes it easy to initialize object to values without having to be concerned with the implementation. For example, on a reset operation we may simply initialize a state machine to the leftmost state of the enumerated list of possible states, that is, statetype'**left**.

TABLE 4-2 Some useful value attributes

Value attribute	Value
scalar_name'**left**	returns the left most value of scalar_name in its defined range
scalar_name'**right**	returns the right most value of scalar_name in its defined range
scalar_name'**high**	returns the highest value of scalar_name in its range
scalar_name'**low**	returns the lowest value of scalar_name in its range
scalar_name'**ascending**	returns true if scalar_name has an ascending range of values
array_name'**length**	returns the number of elements in the array array_name

Finally, a very useful attribute of arrays is the **range** attribute. This is useful in writing loops. For example, consider a loop that scans all of the elements in an array value_array(). The index range is returned by value_array'**range**. This makes it very easy to write loops, particularly when we may not know the size of the array, as is sometimes the case when writing functions or procedures where the array size is determined when the function is called. When we do not know the array size, we can write the loop as shown below:

for i **in** value_array'**range loop**
...
my_var := value_array(i);
...
end loop;

Some specific examples of the use of the range attribute can be found in the discussion of functions and procedures in Chapter 6.

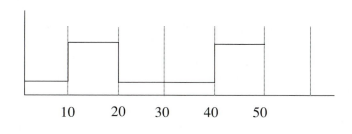

```
library IEEE;
use IEEE.std_logic_1164.all;

entity periodic is
port (Z : out std_logic);
end periodic;

architecture behavioral of periodic is
begin
process
begin
Z <= '0', '1' after 10 ns, '0' after 20 ns, '1' after 40 ns;
wait for 50 ns;
end process;
end behavioral;
```

FIGURE 4-12 An example of the generation of periodic waveforms

4.6 Generating Clocks and Periodic Waveforms

Since wait statements provide the programmer with explicit control over the reactivation of processes, they can be used for generating periodic waveforms as shown in the following example.

Example: Generating Periodic Waveforms

We know that in a signal assignment statement we can specify several future events. For example, we might have the following signal assignment statement:

signal <= '0', '1' **after 10 ns**, '0' **after 20 ns**, '1' **after 40 ns**;

The execution of this statement will create the waveform shown in Figure 4-12. Now, if we place this statement within a process and use a wait statement, we can cause the process to be executed repeatedly, producing a periodic waveform. Recall that upon initialization of the VHDL model all processes are executed. Therefore, every process is executed at least once. During initialization the first set of events shown in the above waveform will be produced. The execution of the **wait for** statement causes the process to be reactivated after 50 ns. This will cause the process to be executed again, generating events in the interval 50–100 ns. The process again suspends for 50 ns and the cycle is repeated. By altering the time durations in the statement in the above process, one can envision the generation of many different types of periodic waveforms. For example, if we wished to generate two-phase non-overlapping clocks we could utilize the same approach as shown in the next example.

Example End: Generating Periodic Waveforms

Example: Generating a Two-Phase Clock

An example of a model for the generation of non-overlapping clocks and reset pulses is shown in Figure 4-13. Such signals are very useful and found in the majority of circuits we will come across. The reset process is a single concurrent signal assignment statement and therefore we can dispense with the **begin** and **end** statements. Recall that CSAs are processes and we can assign them labels. Every process is executed just once, at initialization. During this initialization reset is executed, generating a pulse of width 10 ns. Since there are no input signals or wait statements, the reset process is never executed again! The clock process, on the other hand, generates multiple clock edges in a 20 ns interval with each statement. Note the width of the pulses in the second clock signal: it is adjusted so as to prevent the pulses from overlapping. The **wait for** statement causes the process to be executed again 20 ns later, when each statement generates clock edges in the next 20 ns interval. This process repeats indefinitely, generating the waveforms shown in Figure 4-13. Note how concurrent signal assignment statements are mixed in with the process construct. This type of modeling using both concurrent and sequential statements, is quite common. This is not surprising, since CSAs are essentially processes. If we take the viewpoint that all statements in VHDL are concurrent, then processes using sequential statements can be viewed as one complex signal assignment statement. This is a useful template to have in mind when constructing behavioral models of hardware.

Example End: Generating a Two-Phase Clock

```
library IEEE;
use IEEE.std_logic_1164.all;
entity two_phase is
port (phi1, phi2, reset : out std_logic);
end two_phase;

architecture behavioral of two_phase is
begin
```

```
reset_process: reset <= '1', '0' after 10 ns;
```
Process to generate a reset pulse

```
clock_process: process
begin
phi1 <= '1', '0' after 10 ns;
phi2 <= '0', '1' after 12 ns, '0' after 18 ns;
wait for 20 ns;
```
Clock process

```
end process clock_process;
end behavioral;
```

reset

phi1

phi2

10   20   30   40   50   60
Time (ns)

Events specified by the clock
and reset processes

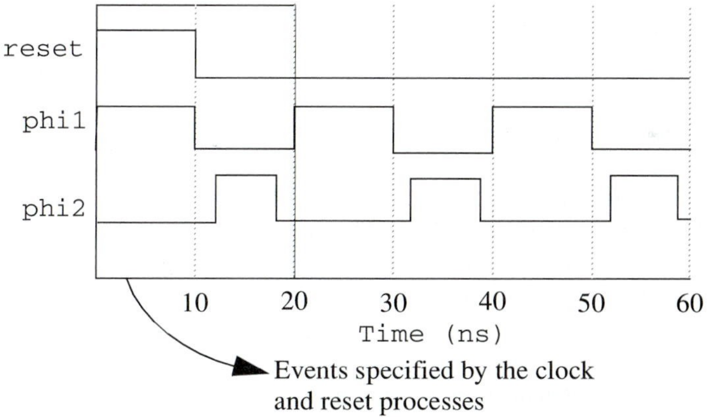

FIGURE 4-13 Generation of two-phase non-overlapping clocks

4.7 Using Signals in a Process

We can think of processes as conventional programs that can be used in VHDL simulation models to provide us with powerful techniques for the computation of events. However, the sequential nature of processes in conjunction with the use of signals within a process can produce behavior that may be unexpected. For example, consider the circuit and corre-

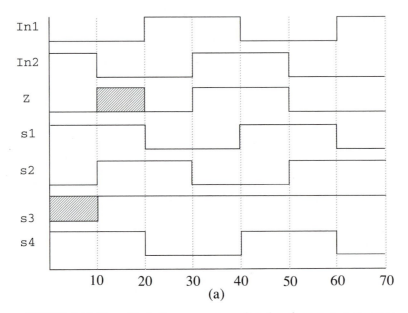

FIGURE 4-14 The effect of processes on signal assignment statements

sponding model and timing behavior illustrated in Figure 3-12 and Figure 3-13. Now, let us enclose the concurrent signal assignment statements shown in the model of Figure 3-12 in a process. This would represent a different model of operation for the following reason: the circuit being described is a combinational circuit; signal values are determined in a data-driven manner; signal assignment statements are executed only when input signal values change. Thus, the value of signal s3 is computed only when there is a transaction on signals s1 or In2.

Now, consider the implementation using processes for which the timing is shown in Figure 4-14. When there is a change in value on either In1 or In2, the process is executed. By definition, a process executes to completion. All statements within a process are executed! Consider the value of signal s3 at initialization time when each process is executed. The value of this signal is undefined. Signal s4 has a value 1 since In1 is 0. Therefore the value of s4 is 1 regardless of the value of s2. However, the value of s3 is undefined and it appears as such on the trace. Now, the process suspends and waits for an event on In1 or In2. It will not be executed again until 10 ns later, when a $1 \rightarrow 0$ transition occurs on In2. From the trace in Figure 4-14 we can see that forcing the process statements to be executed in order and being sensitive only to the inputs In1 and In2 produces a trace very distinct from that in Figure 3-13. Although both models were intended to represent the same circuit, we have realized different behaviors. One could make a good case that the use of a process places artificial constraints on the evaluation of signal values

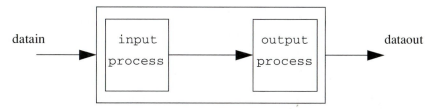

FIGURE 4-15 An example of asynchronous communication

and that the former model is indeed a more accurate reflection of the physical behavior of the circuit. In reality, if we made the process sensitive to all of the signals (i.e., including s1, s2, s3, and s4) we would find that the process model would behave exactly as the earlier model producing an identical trace. However, this is not a very intuitive description of the model. Suffice to say that when using signals within a process, one must be careful that the behavior that results is indeed what the modeler had in mind.

Simulation Exercise 4.2: Use of Wait Statements

This simulation exercise provides an introduction to the use of wait statements to suspend and resume processing under programmer control. We can use such constructs to respond to asynchronous external events. Figure 4-15 shows an example of a simple interface that reads data from an external device and buffers it for an output device. Let us model this interface with two processes. The first process communicates with the second via a handshaking protocol such as the one demonstrated in Figure 4-10. The second process can then drive an external device such as a display.

Step 1. Using a text editor, construct a VHDL model for communication between an input process and an output process using the handshaking protocol captured in Figure 4-10. Assume that the input process can read only a single word at a time. The input process receives a single 32-bit word comprised of four bytes. This word is to be transferred to an output device whose storage layout requires reversing the byte order within the word. This reversal is performed by the input process before it is transferred to the output process, which in turn writes the value to an output port.

Step 2. Use the types std_logic and std_logic_vector for input and output signals. Declare and reference the library IEEE and package std_logic_1164.

Step 3. Create a sequence of 32-bit words as test inputs.

Step 4. Compile the model and load into the simulator.

Step 5. Assign a delay of 10 ns for each handshake transition.

Step 6. Open a trace window and select the signals to be traced. Since we are dealing with 32-bit quantities, set the trace to display these signal values in hexadecimal notation. This will make it considerably easier to read the values in the trace. Such commands are simulator specific.

Step 7. Simulate for several hundred nanoseconds.

Step 8. Trace the four-phase handshake sequence of Figure 4-10.

Step 9. From the trace determine how long it takes for the input process to transfer a data item to the output process.

Step 10. Does the rate at which the input items are provided matter? What happens as we increase the frequency with which data items are presented to the input process?

End Simulation Exercise 4.2

4.8 Modeling State Machines

The examples that have been discussed so far were combinational and sequential circuits in isolation. Processes that model combinational circuits are sensitive to the inputs, being activated whenever an event occurs on an input signal. In contrast, sequential circuits retain information stored in internal devices such as flip-flops and latches. The values stored in these devices are referred to as the *state* of the circuit. The values of the output signals may now be computed as functions of the internal state and values of the input signals. The values of the state variables may also change as a function of the input signals, but are generally updated at discrete points in time determined by a periodic signal such as the clock. Provided with a finite number of storage elements, the number of unique states is finite and such circuits are referred to as finite state machines.

Figure 4-16 shows a general model of a finite state machine. The circuit consists of a combinational component and a sequential component. The sequential component consists of memory elements, such as edge-triggered flip-flops, that record the state and are updated synchronously on the rising edge of the clock signal. The combinational component is comprised of logic gates that compute two Boolean functions. The *output function* computes the values of the output signals. The *next state function* computes the new values of the memory elements (i.e., the value of the next state).

Figure 4-16 suggests a very natural VHDL implementation using communicating concurrent processes. The combinational component can be implemented within one process. This process is sensitive to events on the input signals and the state. Thus, if any of the input signals or the state variables change value this process is executed to compute new values for the output signals and the new state variables. The sequential component can be implemented within a second process. This process is sensitive to the rising edge of the clock signal. When it is executed, the state variables are updated to reflect the value of the next state computed by the combinational component. The VHDL description of such

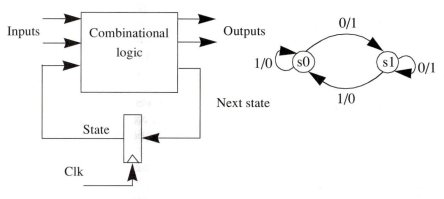

FIGURE 4-16 A behavioral model of a finite state machine

a state machine is shown in Figure 4-17. The model is structured as two communicating processes, with the signals state and next_state used to communicate values between them. The structure of the process comb_process, representing the combinational component, is very intuitive. This process is constructed using a CASE statement. Each branch of the case represents one of the states and includes the output function and next-state function, as shown. The process clk_process updates the state variable on the rising edge of the clock. On reset, this process initializes the state machine to state state0.

There are several interesting aspects to this model. First, note the use of enumerated types for the definition of a state. The model includes the definition of a new type referred to as statetype. This type can take on the values state0 and state1. This is referred to as an *enumerated type*, since we have enumerated all possible values that a signal of this type can take: in this case exactly two distinct values, state0 and state1. This enables much more readable and intuitive VHDL code. The case statement essentially describes the state machine diagram in Figure 4-16. In the clock process, note how the initial state is initialized on reset by using attributes discussed in Section 4.5. The clause statetype'**left** will return the value at the leftmost value of the enumeration of the possible values for statetype. Therefore the initialization shown in the declaration is really not necessary. This is a common form of initialization and is also defined for other types. The default initialization value for signals is signal_name'**left**. A simulation of this state machine would produce a trace of the behavior, as shown in Figure 4-18. Note how the state labels appear in the trace, making it easier to read. Also note that the signal next_state is changing with the input signal, whereas the signal state is not. This is because state is updated only on the rising clock edge, while next_state changes whenever the input signal, x, changes.

The above example could just as easily have been written as a single process whose execution is initiated by the clock edge. In this case, the computation of the next state, the

```vhdl
library IEEE;
use  IEEE.std_logic_1164.all;
entity state_machine is
port(reset, clk, x : in std_logic;
     z : out std_logic);
end  state_machine;

architecture behavioral of state_machine is
type statetype is (state0, state1);
signal state, next_state : statetype := state0;
begin
comb_process: process (state, x)
begin
 case state is                              -- depending upon the current state
 when state0 =>                             -- set output signals and next state
                 if x = '0' then
                 next_state <= state1;
                 z  <= '1';
          else   next_state <= state0;
                 z  <= '0';
          end if;
 when state1 =>
 if x = '1' then
                 next_state <= state0;
                 z <= '0';
          else   next_state <= state1;
                 z  <= '1';
          end if;
 end case;
end process comb_process;

clk_process: process
begin
wait until (clk'event and clk = '1'); -- wait until the rising edge
             if reset = '1' then     -- check for reset and initialize state
             state <= statetype'left;
             else   state <= next_state;
             end if;
end process clk_process;
end behavioral;
```

FIGURE 4-17 Implementation of the state machine in Figure 4-16

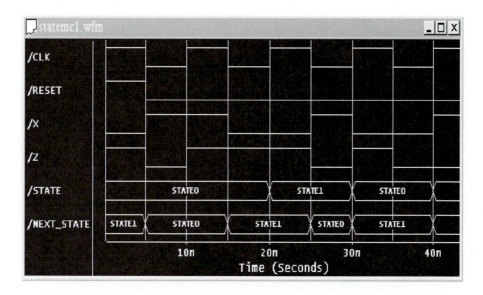

FIGURE 4-18 A trace of the operation of the state machine in Figure 4-17

output signals, and the state transition are all synchronized by the clock. Alternatively, we could construct a model where outputs are computed asynchronously with the computation of the next state. Such a model is shown in Figure 4-19. This model is constructed with three processes: one each for the output function, next-state function, and the state transition. Note how the model is constructed from the structure of the hardware: concurrency in the circuit naturally appears as multiple concurrent processes in the VHDL model. State machines wherein the output signal values are computed only as a function of the current state are referred to as Moore machines. State machines where the output values are computed as a function of both the current state and the input values are Mealy machines. It is evident that the above approaches to constructing state machines enable the construction of both Moore and Mealy machines by appropriately coding the output function.

4.9 Constructing VHDL Models Using Processes

A prescription for writing VHDL models using processes can now be provided. The first step is the same as that described in Section 3.4 for constructing   models using CSAs: we construct a fully annotated schematic of the system being modeled. Figure 4-20 illustrates

```vhdl
library IEEE;
use  IEEE.std_logic_1164.all;
entity state_machine is
port(reset, clk, x : in std_logic;
     z : out std_logic);
end  state_machine;

architecture behavioral of state_machine is
type statetype is (state0, state1);
signal state,  next_state :statetype :=state0;
begin
output_process: process (state, x)
begin
 case state is          -- depending upon the current state
when state0 =>          -- set output signals and next state
             if x = '1' then z <= '0';
             else z <= '1';
             end if;
when state1 =>
             if x = '1' then z <= '0';
             else z <= '1';
             end if;
end case;
end process output_process;
next_state_process: process (state, x)
begin
 case state is          -- depending upon the current state
when state0 =>          -- set output signals and next state
             if x = '1' then next_state <= state0;
             else next_state <= state1;
             end if;
when state1 =>
             if x = '1' then next_state <= state0;
             else next_state <= state1;
             end if;
end case;
end process next_state_process;

clk_process: process
begin
wait until (clk'event and clk = '1');   -- wait until the rising edge
```

FIGURE 4-19 Alternative model for a finite state machine

```
if reset = '1' then  state <= statetype'left;
else state <= next_state;
end if;
end process clk_process;
end behavioral;
```

FIGURE 4-19 (cont.)

a template for constructing such VHDL behavioral models. One approach for translating the annotated schematic to a VHDL model described in the template of Figure 4-20 is as follows:

Construct_ Process_Model

1. At this point I recommend using the IEEE 1164 value system. To do so, include the following two lines at the top of your model declaration:

   ```
   library IEEE;
   use IEEE.std_logic_1164.all;
   ```

   Single-bit signals can be declared to be of type std_logic, while multibit quantities can be declared to be of type std_logic_vector.

2. Select a name for the entity (entity_name) and write the entity description specifying each input and output signal port, its mode, and associated type.

3. Select a name for the architecture (arch_name) and write the architecture description. Place both the entity and architecture descriptions in the same file (as we will see in Chapter 8, this is not necessary in general).

   3.1 Within the architecture description name and declare all of the internal signals used to connect the components. The architecture declaration states the type of each signal, and may include initialization. The information you need is available from your fully annotated schematic.

   3.2 Each internal signal is driven by exactly one component. The computation of values on each internal signal can be described using a CSA or a process.

Using CSAs

For each internal signal select a concurrent signal assignment statement that expresses the value of this internal signal as a function of the signals that are inputs to that component. Use the value of the propagation delay through the component provided for that output signal.

library library-name-1, library-name-2;

use library-name-1.package-name.all;

use library-name-2.package-name.all;

entity entity_name **is**

port(*input signals* : **in** *type*;

    output signals : **out** *type*);

end entity_name;

architecture arch_name **of** entity_name **is**

-- declare internal signals, you may have multiple signals of
-- different types

signal *internal signals* : *type* := initialization;

begin

```
label-1: process(-- sensitivity list --)
--- declare variables to be used in the process
variable variable_names : type:= initialization;
       begin
-- process body
end process label-1;
```
First
Process

```
label-2: process
--- declare variables to be used in the process
variable variable_names : type:= initialization;
       begin
wait until (-- predicate--);
-- sequential statements
wait until (-- predicate--);
-- sequential statements
end process label-2;
```
Second
Process

internal-signal *or* ports <= *simple, conditional, or selected CSA*

-- other processes or CSAs

end arch_name;

FIGURE 4-20 A template for a VHDL model using CSAs and processes

Using a Process

Alternatively, if the computation of the signal values at the outputs of the component are too complex to represent using concurrent signal assignment statements, describe the behavior of the component with a process. One process can be used to compute the values of all of the output signals from that component.

3.2.1 Label the process. If you are using a sensitivity list, identify the signals that will activate the process and place them in the sensitivity list.

3.2.2 Declare variables used within the process.

3.2.3 Write the body of the process, computing the values of output signals and the relative time at which these output signals assume these values. These output signals may be internal to the architecture or may be port signals found in the entity description. If a sensitivity list is not used, specify wait statements at appropriate points in the process to specify when the process should suspend and when it should resume execution. It is an error to have both a sensitivity list and a wait statement within the process.

3.3 If there are signals that are driven by more than one source, the type of this signal must be a resolved type. This type must have a resolution function declared for use with signals of this type. For our purposes, use the IEEE 1164 types `std_logic` for single-bit signals and `std_logic_vector` for bytes, words, or multibit quantities. These are resolved types. Make sure you include the **library** clause and the **use** clause to include all of the definitions provided in the `std_logic_1164` package.

3.4 If you are using any functions or type definitions provided by a third party, make sure that you have declared the appropriate library using the **library** clause and declared the use of this package via the presence of a **use** clause in your model.

The behavioral model will now appear structurally as shown in Figure 4-20, inclusive of both CSA and sequential statements.

Simulation Exercise 4.3: State Machines

Consider the state machine shown in Figure 4-21. This state machine has two inputs. The first is the `reset` signal, which initializes the machine to state 0. The second is a bit-serial input. The state machine is designed to recognize the sequence 101 in the input sequence and set the value of the output to 1. The value of the output remains at 1 until the state machine is reset.

Step 1. Write the VHDL model for this state machine. Use an enumerated type to represent the state (i.e., the type `statetype` as in Figure 4-19). Structure your state machine description as three processes.

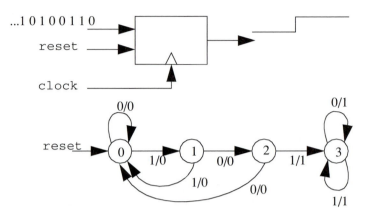

FIGURE 4-21 State machine for recognizing bit patterns

Process 1: Write an output process that determines the value of the single-bit output based on the current state and the value of the single-bit input.

Process 2: Write a process that computes the next state based on the value of the input signal and the current state.

Process 3: Write a clock process that updates the state on the rising edge of the clock signal. On a `reset` pulse the state machine is reset. Otherwise the state is modified to reflect the next state

Step 2. Within the simulator that you are using, structure the input stimulus as follows.

Apply a clock signal with a period of 20 ns.

Apply a reset signal that generates a single pulse of duration 30 ns.

Generate a random, bit-serial sequence with the pattern 101 embedded within the sequence.

Step 3. Simulate the model long enough to detect the pattern in the input sequence.

Step 4. Modify the model to recognize other patterns and repeat.

Step 5. Modify the model so that after the pattern is recognized the state machine is reinitialized to state 0. Detection of the pattern now results in a pulse on the output signal.

End Simulation Exercise 4.3

4.10 Common Programming Errors

The following are some common programming errors that are made during the learning process.

4.10.1 Common Syntax Errors

- Do not forget the semicolon at the end of a statement.

- Remember it is **elsif** and not **elseif** !

- It is an error to use **endif** instead of **end if.**

- Do not forget to leave a space between the number and the time base designation. For example, it should read 10 **ns** and not 10**ns**.

- Use underscore and not hyphens in label names. Thus, not `half-adder` but `half_adder`.

- Expressions on the right-hand side of an assignment statement may have binary numbers expressed as x"00000000". When we use the IEEE 1164 types, even though `std_logic_vector` is a vector of bit signals, they are not the same type as x"00000000". Simulators may require you to perform a type conversion operation to convert the type of this binary number to `std_logic_vector` before you can make this assignment. The function `to_stdlogicvector`(x"00000000") is available in the package `std_logic_1164` that is in the library `IEEE`. Check the documentation for the VHDL simulator you are using. Such type mismatches will be caught by the compiler.

4.10.2 Common Run-Time Errors

- It is not uncommon to sometimes use signals when you should use variables. Signals will be updated only after the next simulation cycle. Thus, you will find that values take effect later than you expected, for example, one clock cycle later.

- If you use more than one process to drive a signal, the value of the signal may be undefined unless you use resolved types and have specified a resolution function. Make sure there is only one source (e.g., process) for a signal unless you mean to have shared signals. When using the `std_logic_1164` package, use `std_logic` and `std_logic_vector` types. These are resolved types that provide an associated resolution function.

- A process should have a sensitivity list or a wait statement.

- A process cannot have both a sensitivity list and a wait statement.

- Remember that all processes will be executed once when the simulation is started. This can sometimes cause unintended side effects if your processes are not explicitly controlled by wait statements.

4.11 Chapter Summary

This chapter has introduced models that use processes and the use of sequential statements. This is a generalization of the behavioral models with concurrent signal assignment statements described in Chapter 3. The concepts introduced in this chapter include.

- Processes
- Sequential statements
    - if-then-else
    - case
    - loop
- Wait statements
- Attributes
- Communicating processes
- Modeling state machines
- Using both CSAs and processes within the same architecture description

Exercises

1. Construct a VHDL model of a parity generator for 7-bit words. The parity bit is generated to create an even number of bits in the word with a value of 1. Do not prescribe propagation delays to any of the components. Simulate the model and check for functional correctness.

2. Explain why you cannot have both a sensitivity list and wait statements within a process.

3. Construct and test a model of a negative edge-triggered JK flip-flop.

4. Consider the construction of a register file with 8 registers, where each register is 32 bits. Implement the model with two processes. One process reads the register file, while another writes the register file. You can implement the registers as signals declared within the architecture and therefore visible to each process.

5. Implement a 32-bit ALU with support for the following operations: add, sub, and, or, and complement. The ALU should also produce an output signal that is asserted when the ALU output is 0. This signal may be used to implement branch instructions in a processor datapath.

6. Show an example of VHDL code that transforms an input periodic clock signal to an output signal at half the frequency.

7. Construct a VHDL model for generating four-phase non-overlapping clock signals. Pick your own parameters for pulse width and pulse separation intervals.

8. Implement and test a 16-bit up–down counter.

9. Implement and test a VHDL model for the state machine for a traffic-light controller [4] shown in Figure 4-22.

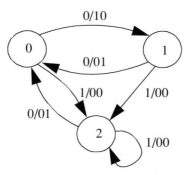

FIGURE 4-22 State machine for a traffic-light controller

10. Consider a variant of Simulation Exercise 4.3 where we are interested in the occurrence of six 1's in the bit stream. After six 1's have been detected, the output remains asserted until the state machine is reset. Construct and test this model.

CHAPTER 5

Modeling Structure

Our model of a digital system remains that of an interconnected set of components. The preceding chapters described how the behavior of each component could be specified in VHDL. Informally, the behavior of each component is specified as the set of output events that occur in response to input events. Behavioral descriptions of a component may be specified using concurrent signal assignment statements. When this is infeasible due to the complexity of the event calculations, the behavioral models are specified using one or more processes and sequential statements. A third approach to describing a system is simply in terms of the interconnection of its components. Rather than focusing on what each component does we are concerned with simply describing how components are connected. Behavioral models of each component are assumed to be provided. Such a description is referred to as a *structural model*. Such models describe only the structure of a system, without regard to the operation of individual components. This chapter discusses the construction of structural models in VHDL.

5.1 Describing Structure

A common means of conveying structural descriptions is through block diagrams. Components represented by blocks are interconnected by lines representing signals. In Chapter 3, models of a full adder using concurrent signal assignment statements were described. These models provide a description of "what" the system does. Instead of employing such a model suppose we wish to describe the circuit as being constructed from two half adders. Such a design is shown in Figure 5-1. In understanding how such a design may be described in VHDL, imagine conveying this schematic over the telephone (no faxes!) to a friend who has no knowledge of full-adder circuits. You would like to have them correctly reproduce the schematic as you describe it. You can also think of describing this schematic across the table to someone without showing them the diagram or taking a pen to paper yourself. Do not use your fingers or hands to point and use only verbal guidelines! When we think of conveying descriptions in these terms we see that we need precise and unambiguous ways to describe structure.

We might convey such a description as follows. First, we must describe the inputs and outputs to the full adder. This is not too difficult to convey verbally. We can also easily state the type and mode of the input–output signals, for example, whether they are single-bit signals, input signals, or output signals. This information constitutes the corresponding entity description. Now imagine describing the interconnection of the components over the telephone. You would probably first list the components you need: two half adders and a two-input OR gate. Conveying such a list of components verbally is also not difficult, but now comes the tricky part. How do you describe the interconnection of these components unambiguously? In order to do so you must first be able to distinguish between components of the same type. For example, in Figure 5-1(a), the half adders must be distinguished by assigning them unique labels, such as H1 and H2. The signals that will be used to connect these components are also similarly labeled. For example, we may label them s1, s2, and s3. Additional annotations that we need are the labels for the ports in a half adder and the ports in an OR gate. Such detailed annotations are necessary so that we can refer to them unambiguously, such as the sum output port of half-adder H1 or the sum output port of half adder H2. We have now completed the schematic annotation to a point where the circuit can be described in a manner that allows the person at the other end of the telephone to draw it correctly. For example, the interconnection between H1 and H2 can be described by stating that the sum output of H1 is connected to signal s1, and the a input of H2 is connected to signal s1. Implicitly we have stated that the sum output of H1 is connected to the a input of H2 using signal s1. We have done this indirectly by describing connections between the ports and signal s1. There is an analogy with the physical process of constructing this circuit. If you were wiring this circuit on a protoboard in the laboratory you would actually use signal wires to connect components. Once all of the components and their input and output ports are labeled we see that we can describe this circuit in a manner that will allow the person to correctly build the circuit.

Based on the above example we can identify a number of features that a formal, VHDL structural description might possess: (i) the ability to define the list of components, (ii) the definition of a set of signals to be used to interconnect these components, and (iii) the ability to uniquely label, and therefore distinguish, between multiple copies of the same component. The VHDL syntax that realizes these features in an architecture description is shown in Figure 5-2.

The component declaration includes a list of the components being used and the input and output signals of each component; or, in VHDL terminology, the input and output ports of the component. This declaration essentially states that this architecture will be

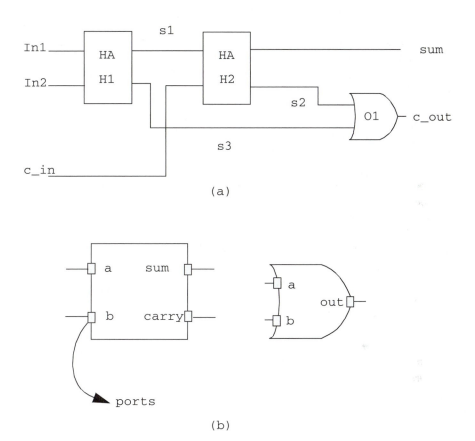

FIGURE 5-1 (a) Full-adder circuit (b) Interface description of the half-adder and OR-gate component

using components named half_adder and or_2 gate, but so far has not stated how many of each type of component will be used. The declaration of the components is followed by the declaration of all of the signals that will be used to interconnect the components. These signals would correspond to the set of signal wires you might use in the laboratory. From a programming languages point of view we note similarities with the manner in which we construct Pascal or C programs. In a Pascal program we declare the variables and data structures that we will use (e.g., arrays) and their types before we actually use them. In a VHDL structural model as shown in Figure 5-2, we declare all of the components and signals that we will use before we describe how they are interconnected. Collectively, the component and signal declarations complete the declarative part of the **architecture** construct. This is analogous to the parts list that you would have if you were

```
library IEEE;
use IEEE.std_logic_1164.all;
entity full_adder is
port (In1, In2, c_in : in std_logic;
     sum, c_out : out std_logic);
end full_adder;

architecture structural of full_adder is
component half_adder
port (a, b: in std_logic;                         ⎤
     sum, carry: out std_logic);                   │
end component;                                      │   Component
                                                    │   Declarations
component or_2                                      │
 port (a, b : in std_logic;                         │
       c : out std_logic);                          │
end component;                                      ⎦

                                                   ⎤  Signal
signal s1, s2, s3 : std_logic;                     ⎦  Declarations
begin
H1: half_adder port map(a => In1, b => In2,        ⎤
                  sum=>s1,carry=>s3);               │
H2:  half_adder port map(a => s1, b => c_in,        │   Component
              sum => sum, carry => s2);             │   Interconnections
O1:  or_2 port map(a => s2, b => s3,                │
              c => c_out);                          ⎦

end structural;
```

FIGURE 5-2 Structural model of a full adder

to build this circuit in the laboratory. Now all that remains is the process of actually "wiring" the components together. This is provided in the **architecture** body that follows the declarative part, and is delimited by the **begin** and **end** statements.

Consider the first statement in the architecture body. Let us go back to our analogy of wiring the circuit on a protoboard in the laboratory. We must first acquire the components, label them, and lay them out on the board. Each component must then be connected to other components or circuit inputs using the signal wires. Each line in the architecture body provides this information for each component. This is a component *instantiation* statement.

Recall that processes and concurrent signal assignments can be labeled. Components are similarly labeled. The first word, H1, is the label of a half-adder component. This is followed by the **port map**() construct. This construct states how the input and output ports of H1 are connected to other signals and ports. The first argument of the port map construct simply states that the a input port of H1 is connected to the In1 port of the entity full_adder. The third argument states that the sum output port of H1 is connected to the s1 signal. The remaining port map constructs can be similarly interpreted. Note that both components have identical input port names. This is not a problem since the name is associated with a specific component which is unambiguously labeled. It is as if we were peering at the circuit through a keyhole and could see only one component at a time. We must completely describe the connections of all of the ports of each component before we can describe those of the next component. It is apparent that the interconnection of the schematic in Figure 5-1(a) is completely specified in the structural description of Figure 5-2.

Finally, one other important feature of this model should be noted. The behavioral models of each component are assumed to be provided elsewhere, that is, there are entity–architecture pairs describing a half adder. Note that there are no implications on the type of model used to describe the operation of the half adder. This behavioral model could comprise concurrent signal assignment statements, or use processes and sequential statements, or itself be a structural model describing a half adder as the interconnection of gate level behavioral models. Such hierarchies are very useful and are discussed in greater detail later in this chapter. The structural model shown in Figure 5-2 only states that a half-adder description, whose entity is labeled half_adder, is to be used.

Example: Structural Model of a State Machine

Consider a bit-serial adder shown in Figure 5-3. Two operands are applied serially, bit by bit, to the two inputs. On successive clock cycles, the combinational logic component computes the sum and carry values for each bit position. The D flip-flop stores the carry bit between additions of successive bits and is initialized to 0. As successive bits are added the corresponding bits of the output value are produced serially on the output signal. The state machine diagram is shown below the circuit. As described in Section 4.8, we can develop this as a behavioral model of a state machine that implements the state diagram shown in the figure. In this example we show the structural model of the state machine implementation. We have redrawn this circuit in the manner shown in Chapter 4 with a combinational logic component and a sequential logic component. The corresponding VHDL structural model is shown in Figure 5-4. Note the use of the **open** clause for component labeled D1. The signal qbar at the output of the flip-flop is not used. The **port**

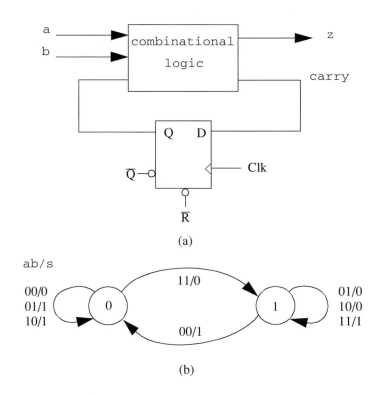

FIGURE 5-3 A bit-serial adder (a) Logic implementation (b) State diagram

map() clause in the component instantiation statement expresses this by setting the signal to **open**. The hardware analogy is that of an output pin on a chip that remains unconnected.

Example End: Structural Model of a State Machine

5.2 Constructing Structural VHDL Models

We are now ready to provide a prescription for constructing structural VHDL models. As with behavioral models, this simple methodology comprises two steps: (i) the drawing of an annotated schematic, and (ii) the conversion to a VHDL model.

```
library IEEE;
use IEEE.std_logic_1164.all;
entity serial_adder is
port (a, b, clk, reset : in std_logic;
      z : out std_logic);
end serial_adder;

architecture structural of serial_adder is
--
-- declare the components that we will be using
--
component comb
 port (a, b, c_in : in std_logic;
       z, carry : out std_logic);
end component;
component dff
port (clk, reset, d : in std_logic;
        q, qbar : out std_logic);
end component;
signal s1, s2 :std_logic;
begin
--
-- describe the component interconnection
--
C1: comb  port map (a => a, b => b, c_in => s1, z =>z, carry => s2);
D1: dff  port map(clk => clk, reset =>reset, d=> s2, q=>s1,
                  qbar => open);
end structural;
```

FIGURE 5-4 VHDL structural model of the bit-serial adder

Construct_Structural_Schematic

1. Ensure that you have a behavioral or structural description of each component in the system being modeled. This means that you have a correct, working entity–architecture description of each component. Using the entity descriptions, create a block for each component with the input and output ports labeled.

2. Connect each port of each component to the port of another entity, or to an input or output port of the system being modeled.

3. Label each component with a unique identifier: H1, U2, and so on.

4. Label each internal signal with a unique signal name and associate a type with this signal, for example, `std_logic_vector`. Make sure all of the ports connected to an internal signal, as well as the signal are of the same type!

5. Label each system input port and output port and define its mode and type.

This annotated schematic can be transcribed into a structural VHDL model. Figure 5-5 illustrates a template for writing structural models in VHDL. One approach to filling in this template is described in the following procedure. This procedure relies on the availability of the annotated schematic.

Construct_Structural_Model

1. At this point we recommend using the IEEE 1164 value system. To do so include the following two lines at the top of your model declaration.

   library IEEE;
   use IEEE.std_logic_1164.all;

   Single-bit signals can be declared to be of type `std_logic` while multibit quantities can be declared to be of type `std_logic_vector`.

2. Select a name for the entity (`entity_name`) representing the system being modeled and write the entity description. Specify each input and output signal port, its mode, and associated type.

3. Select a name for the architecture (`arch_name`) and write the architecture description as follows.

   3.1 Construct one component declaration for each unique component that will be used in the model. A component declaration can be easily constructed from the component's entity description.

   3.2 Within the declarative region of the architecture description—before the **begin** statement—list the component declarations.

   3.3 Following the component declarations name and declare all of the internal signals used to connect the components. These signal names are shown on your schematic. The declaration states the type of each signal and may also provide an initial value.

   3.4 Now we can start writing the architecture body. For each block in your schematic write a component instantiation statement using the **port map**() construct. The component label is derived from the schematic. The **port map**() construct will have as many entries as there are ports on the component. Each element of the port map construct has the form

   port-signal => (internal signal or entity-port)

```
library library-name-1, library-name-2;
use library-name-1.package-name.all;
use library-name-2.package-name.all;
```

entity entity_name **is**
port(*input signals* : **in** *type*;
    output signals : **out** *type*);
end entity_name;

architecture arch_name **of** entity_name **is**

    -- declare components used

component component1_name
port(*input signals* : **in** *type*;
    output signals : **out** *type*);
end component;

 component component2_name
port(*input signals* : **in** *type*;
    output signals : **out** *type*);
end component;

    -- declare all signals used to connect the components

signal *internal signals* : *type* := *initialization*;

begin

    -- label each component and connect its ports to signals or other ports

Label1: component1-name **port map** (port=> signal,.....);

Label2: component2-name **port map** (port => signal,.....);

end arch_name;

FIGURE 5-5 Structural model template

Each port of the component is connected to an internal signal or to a port of the top-level entity. Remember, the mode and type of the port of the component must match that of the internal signal or entity port that it is connected to the component port.

4. Some signals may be driven from more than one source. If this is the case then this signal is a shared signal and its type must be a resolved type. We can either define a new resolved type and its associated resolution function (as described in Chapter 6) or simply use the IEEE 1164 data types `std_logic` and `std_logic_vector` which are resolved types.

   Modern CAD tools can generate such structural models automatically from designs diagrams created with schematic capture tools available in modern CAD environments.

Simulation Exercise 5.1: A Structural Model

The goal of this exercise is to introduce the reader to the construction, testing, and simulation of a simple structural model.

Step 1. Create a text file with the structural model of the full adder shown in Figure 5-2. Let us refer to this file as *full-adder.vhd*.

Step 2. Create a text file with the model of the half adder shown in Figure 3-2. Let us refer to this file as *half-adder.vhd*. Ensure that the entity name for the half adder in this file is the same as the name you have used for the half-adder component declaration in the full-adder structural model. Remember the environment must have some way of being able to find and use the components that you need when you simulate the model of the full adder. Just as in the laboratory, components names are used for the purpose. In fact the file names can be different as long as you consistently name architectures (`arch_name`), entities (`entity_name`), and components (`component_name`).

Some thought will reveal that such an approach follows intuition. Designs must be described in terms of lower level design units. File names are an artifact of the computer system we are using. When we compile an entity named `E1` in a file called *homework1.vhd*, we will see that compiled unit names will be based on the label `E1` rather than *homework1.vhd*. This will enable compilation of higher level structural models to find relevant files based on the component names and not the filenames that they are stored in. More on the programming mechanics in Chapter 8.

Step 3. Create a text file with a model of a 2-input OR gate. Let us refer to this file as *or2.vhd*. Use a gate delay of 5 ns. Again make sure that the entity name is the same as the component name for the 2-input OR gate model declared in the model of the full adder.

Step 4. Compile the files *or2.vhd, half-adder.vhd*, and *full-adder.vhd* in this order.

Step 5. Load the simulation model into the simulator.

Step 6. Open a trace window with the signals you would like to trace. Include internal signals which are signals that are not entity ports in the model.

Step 7. Generate a test case. Apply the stimulus corresponding to the test case to the inputs. Run the simulation for one time step. Examine the output to ensure it is correct.

Step 8. Run the simulation for 50 ns.

Step 9. Check the behavior of the circuit and note the timing on the internal signals with respect to the component delays.

End Simulation Exercise 5.1

5.3 Hierarchy, Abstraction, and Accuracy

The structural model of the full adder shown in Figure 5-2 presumes the presence of models of the half adder. Although this model could be any one of the behavioral models described in Chapter 3 or Chapter 4, the model could also be a structural model itself as shown in Figure 5-6. Thus, we have a hierarchy of models. This hierarchy can be graphically depicted as shown in Figure 5-7. Each box in the figure denotes a VHDL model: an entity–architecture pair. The architecture component of each pair may in turn reference other entity–architecture pairs. At the lowest level of the hierarchy exist architectures comprised of behavioral rather than structural models of the components.

A few interesting observations can be made about the model shown in Figure 5-6. We see that structural models simply describe interconnections. They do not describe any form of behavior. There are no descriptions of how output events are computed in response to input events. How can the simulation be performed? When the structural model is loaded into the simulator, a simulation model is internally created by replacing the components by their behavioral descriptions. If the description of a component is also a structural description (as is the case in the model of Figure 5-2 using the architecture of the half-adder model in Figure 5-6), then the process is repeated for structural models at each level of the hierarchy. This process is continued until all components of the hierarchy have been replaced by behavioral descriptions. The levels of the hierarchy correspond to different levels of detail or *abstraction*. This process is referred to as *flattening* of the hierarchy. We now have a discrete event model that can be simulated. From this point of view, we see that structural models are a way of managing large, complex designs. A modern design may have several million to tens of millions of gates. It is often infeasible to build a single, flat, simulation model at the level of individual gates to test and evaluate the complete design. This may be due to the amount of simulation time required or the amount of mem-

```
architecture structural of half_adder is
component xor2
 port (a, b : in std_logic;
        c : out std_logic);
end component;
component and2
port (a, b : in std_logic;
         c : out std_logic);
end component;
begin
EX1: xor2 port map (a => a, b => b, c => sum);
OR1: and2 port map (a=> a, b=> b, c=> carry);
end structural;
```

FIGURE 5-6 Structural model of a half adder

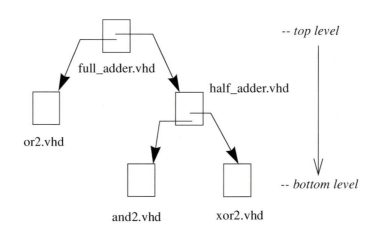

FIGURE 5-7 Hierarchy of models used in the full adder

ory required for such a detailed model. Thus, we may chose to approximate this gate level behavior by constructing less accurate models. For example, we have seen that a state machine may be described at the gate level, or via a process level description (as in Figure 4-16). The latter model is said to be at a higher level of abstraction. The ability to work at multiple levels of abstraction is required to manage large complex designs. For example,

1. We may have a library of VHDL models of distinct components, such as those derived from the manufacturer's component data book. These models have been developed, debugged, and tested. You can construct a model of a circuit by simply using these components. The only model you will have to write is a structural model. You also have to know the component entity description so that you correctly declare the component you are using. This is akin to using a library of mathematical functions in C or Pascal.

2. After three weeks, you might have a new and improved model of a half adder that you would like to use. You can test and debug this model in isolation. Then simply replace the old model with this new model. The need to recompile any dependent design entities is determined by the rules described in Chapter 8.

Finally, we note that the simulation time is directly impacted by the level at which we construct simulation models. Consider the behavioral model of the half adder described in Chapter 4, Figure 4-3. Events on input signals produce events on output signals. In contrast, when we flatten the hierarchy of Figure 5-2, events on input signals will produce events on outputs of the gates within the half adders and eventually propagate to the output signals as events. The more detailed the model, the larger the number of events we must expect the model to generate. The larger the number of events generated by the model the greater the simulation time. As a result, more accurate models will take a significantly longer time to simulate. Generally, the closer we are to making implementation decisions, the more accurate we wish the simulation to be and hence more time is invested in simulation.

Since the full-adder model depends upon the existence of models for the half adder and 2-input OR gate components, it follows that these models must be analyzed before the model for the full adder is analyzed. Once all of the models have been analyzed what happens if we make changes to only some design units? Must we recompile all of the design units each time we make a change to any one? If not, what dependencies between design units must we respect? These issues are discussed in Chapter 8.

Simulation Exercise 5.2: Construction of an 8-bit ALU

The goal of this exercise is to introduce trade-offs in building models at different levels of abstraction and trading accuracy for simulation speed.

Step 1. Start with the model of a single-bit ALU as given in Simulation Exercise 3.2. This model is constructed with concurrent signal assignment statements. Replace this model with one that replaces all of the concurrent signal assignment statements in the architecture body with sequential assignment statements and a single process. The process should be sensitive to events on input signals a, b, and c_in. The process should use variables to compute the value of the ALU output. The last statement in the process should be a signal assignment statement assigning the ALU output value to the signal result. Use a delay of 10 ns through the ALU.

Step 2. Analyze, simulate, and test this model and ensure that all three operations (AND, OR, and ADD) operate correctly.

Step 3. Construct a VHDL structural model of a 4-bit ALU. Use the single-bit ALU as a building block. Use a ripple carry implementation to propagate the carry between single-bit ALUs. Remember to compile the single-bit model before you compile the 4-bit model.

Step 4. Construct an 8-bit ALU using the 4-bit ALU as a building block. Use a ripple carry implementation to propagate the carry signal. Remember to compile the single-bit model before you analyze the 8-bit model.

Step 5. Based on your construction, what is the propagation delay though the 8-bit adder?

Step 6. Open a trace window with the signals you would like to trace. In this case, you will need to trace only the input and output signals to test the model.

Step 7. Generate a test case for each ALU instruction. Apply the stimulus for a test case to the inputs. Run the simulation for a period equal to at least the delay through the 8-bit adder. Examine the output values to ensure it is correct.

Step 8. Print the trace.

Step 9. Rewrite the 8-bit model as a behavioral model rather than a structural model. In this case there is no hierarchy of components. Use a single process and the following hints.

Step 9 (a) Inputs, outputs, and internal variables are all now 8-bit vectors of type std_logic_vector (use the IEEE 1164 value system.)

Step 9 (b) Make use of variables to compute intermediate results.

Step 9 (c) Use the case statement to decode the opcode.

Step 9 (d) Do not forget to set the value of the output carry signal.

Step 9 (e) The propagation delay should be set to the delay through the hierarchical 8-bit model.

Step 10. Test the new model. You should be able to use the same inputs to test this model as you used for the hierarchical model.

Step 11. Generate a trace for the single-level model demonstrating that the model functions correctly.

```
library IEEE;
use IEEE.std_logic_1164.all;

entity xor2 is
generic (gate_delay : Time:= 2 ns);
port(In1, In2 : in std_logic;
     z : out std_logic);
end xor2;

architecture behavioral of xor2 is
begin
z <= (In1 xor In2) after gate_delay;
end behavioral;
```

FIGURE 5-8 An example of the use of generics

Step 12. Qualitatively compare the two models with respect to the difference in the number events that occur in the flattened hierarchical model and the single-level model in response to a new set of inputs.

End Simulation Exercise 5.2

5.4 Generics

It is often useful to be able to parameterize models. For example, a gate-level model may have a parameterized value of the gate delay. The actual value of the gate delay is determined at simulation time by the value that is provided to the model. Having parameterized models makes it possible to construct standardized libraries of models that can be shared. The VHDL language provides the ability to construct parameterized models using concept of *generics*

Figure 5-8 illustrates a parameterized behavioral model of a two-input exclusive-OR gate. The propagation delay in this model is parameterized by the constant gate_delay. The default (or initialized value) value of gate_delay is set to 2 ns. This is the value of delay that will be used in simulation models such as that shown in Figure 5-6, unless a different value is specified. A new value of gate_delay can be specified at the time the model is used, as shown in Figure 5-9. This version of the half adder will make use of exclusive-OR gates that exhibit a propagation delay of 6 ns through the use of the **generic map**() construct. Note the absence of the ';' after the **generic map**() construct!

```
architecture generic_delay of half_adder is
component xor2
generic (gate_delay: Time); -- new value may be specified here instead
port (a, b : in std_logic;      -- of using a generic map() construct
      c : out std_logic);
end component;
component and2
generic (gate_delay: Time);
port (a, b : in std_logic;
      c : out std_logic);
end component;
begin
EX1: xor2 generic map (gate_delay => 6 ns)
          port map(a => a, b => b, c => sum);
A1: and2 generic map (gate_delay => 3 ns)
          port map(a=> a, b=> b, c=> carry);
end generic_delay;
```

FIGURE 5-9 Use of generics in constructing parameterized models

The xor2 model is now quite general. Rather than manually editing and updating the delay values in the VHDL text, we can specify the value we must use when the component is instantiated. Considering the thousands of digital system components that are available, manually modifying each model when we wish to change its attributes can be quite tedious and inefficient.

5.4.1 Specifying Generic Values

The example in Figure 5-9 illustrates how the value of a generic constant can be specified using the **generic map**() construct when the component is instantiated. Alternatively, the value can be specified when the component is declared. For example, in Figure 5-8 the component declaration of xor2 includes a declaration of the generic parameters. This statement can be modified to appear as follows.

generic (gate_delay: **Time**:= 6 ns);

Therefore, we observe that within a structural model there are at least two ways in which the values of generic constants of lower-level components can be specified: (i) in the component declaration, and (ii) in the component instantiation statement using the **generic map**() construct. If both are specified, then the value provided by the **generic**

```
library IEEE;
use IEEE.std_logic_1164.all;

entity half_adder is
generic (gate_delay:Time:= 3 ns);
port (a, b : in std_logic;
     sum, carry : out std_logic);
end half_adder;

architecture generic_delay2 of half_adder is
component xor2
generic (gate_delay: Time);
 port(a, b : in std_logic;
      c : out std_logic);
end component;

component and2
generic (gate_delay: Time);
port (a, b : in std_logic;
     c : out std_logic);
end component;

begin
EX1: xor2 generic map (gate_delay => gate_delay)
          port map(a => a, b => b, c => sum);
A1: and2 generic map (gate_delay => gate_delay)
          port map(a=> a, b=> b, c=> carry);
end generic_delay2;
```

FIGURE 5-10   Passing values of generics through multiple levels of the hierarchy

map() takes precedence. If neither is specified, then the default value defined in the model is used.

The values of these generics can be passed down through multiple levels of the hierarchy. For example, suppose that the full-adder model shown in Figure 5-2 defines the value of gate_delay for all lower level modules. In this case, the half-adder module may be modified to appear as shown in Figure 5-10.

Within the half adder the default gate delay is set to 3 ns. This is the value of gate_delay that is normally passed into the lower level models for xor2 and and2. However, if the full-adder description uses **generic maps** to provide a new value of the gate delay to the half-adder models that are instantiated in the structural description shown in Figure 5-2, then these values take precedence and will flow down to the gate-level models. This process is depicted graphically in Figure 5-11. From the figure it is evident that

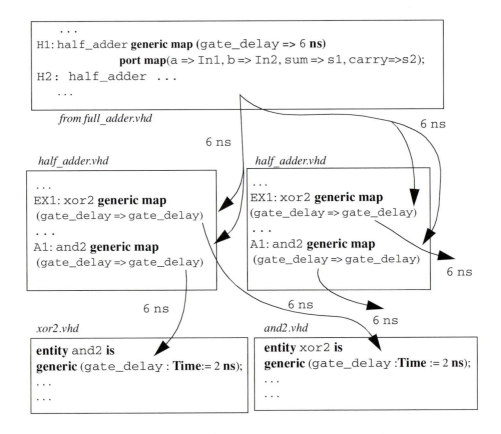

FIGURE 5-11 Parameter passing through the hierarchy using generics

by changing the value of `gate_delay` in the full-adder model, the gate-level VHDL models for `xor2` and `and2` will utilize this value of the gate delay in the simulation. Although the models are written with default gate delay values of 2 ns, the new value overrides this default value.

5.4.2 Some Rules about Using Generics

The terminology is quite appropriate: the use of generics enables us to write *generic* models whose behavior in a particular simulation is determined by the value of the generic parameters. Generics appear very much like ports. They are a part of the interface specification of the component. However, unlike ports, they do not have a physical interpretation. They are more a means of conveying information through the design hierarchy, thereby enabling component designs to be parameterized. Generics are constant objects. Therefore, they cannot be written, but only read. The values of generic parameters must be computable at the time the simulator is loaded with the VHDL model. Therefore, we may include expressions in the value of a generic parameter. However, the value of this expression must be computable at the time the simulator is loaded. Finally, we must be careful about the precedence of the values of generic objects, as described in the preceding sections. The following examples further illustrate the power of constructing parameterized models.

Example: N-Input OR gate

One class of generic gate-level models is one where the number of inputs can be parameterized. Therefore we can have just one VHDL model of an N-input gate. We can produce a 2, 3, or 6 input gate model by setting the value of a generic parameter. This example is shown in Figure 5-12. When this OR-gate model is used in a VHDL model, the generic parameter n must be mapped to the required number of inputs using the **generic map**() construct in a higher level structural model. For example, if we were using several OR gates in a structural model, we would have one instantiation statement for each model. Each instantiation statement would include the following statement: **generic map** (n=>3);

```
library IEEE;
use IEEE.std_logic_1164.all;
entity generic_or is
generic (n : positive:=2);
port (in1 : in std_logic_vector ((n-1) downto 0);
    z : out std_logic);
end generic_or;

architecture behavioral of generic_or is
begin
process (in1)
variable sum : std_logic:= '0';
begin
sum := '0'; -- on an input signal transition sum must be reset to 0
for i in 0 to (n-1) loop
sum := sum or in1(i);
end loop;
z <= sum;
end process;
end behavioral;
```

FIGURE 5-12 An example of a parameterized gate-level model

In this statement the value 3 would be replaced by the number of inputs for that particular gate. Note that there is only one VHDL model. The conventional programming language analogy would be that of using functions where the arguments determine the computation performed by the function.

Example End: N-Input OR gate

Example: N-Bit Register

Another example of the use of generics is a parameterized model of an N-bit register. Generics may be used to configure the model in a specific instance to be of a fixed number of bits. Let us consider a register comprised of D flip-flops with asynchronous reset and load enable signals. The model is shown in Figure 5-13. The operation of the register (process) is sensitive to the occurrence of an event on the reset or clk signals. The size of the register is determined when this model is instantiated by a higher-level model. The

```
library IEEE;
use IEEE.std_logic_1164.all;
entity generic_reg is
generic (n : positive:=2);
port ( clk, reset, enable : in std_logic;
       d : in std_logic_vector (n-1 downto 0);
       q : out std_logic_vector (n-1 downto 0));
end generic_reg;

architecture behavioral of generic_reg is
begin
reg_process: process (clk, reset)
begin
 if reset = '1' then
    q <= (others => '0');
 elsif (clk'event and clk = '1') then
    if enable = '1' then
       q <= d;
    end if;
  end if;
end process reg_process;
end behavioral;
```

FIGURE 5-13 An example of a parameterized model of an N-bit register

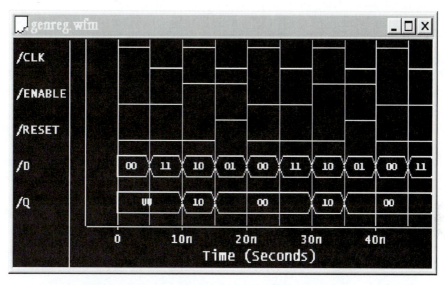

FIGURE 5-14 Trace of the operation of a 2-bit register

default value is a 2-bit register. Note how the value of q is set using the **"others"** construct. Since q is a vector of bits, this statement provides a concise approach to specifying the values of all the bits in a vector when they are equal. A trace of the operation of this 2-bit register in the presence of various waveforms on the input signals is also shown in Figure 5-14.

Example End: N-bit Register

Simulation Exercise 5.3: Use of Generics

This exercise illustrates the utility of the use of generics for parameter passing in structural models.

Step 1. Start with the model of a single-bit ALU that is written using CSAs from Simulation Exercise 3.2. Modify this model to include a generic parameter to specify the gate delay. Set the default gate delay to 3 ns. Use 2 ns for the delay through the multiplexor.

Step 2. Compile, simulate, and test this model and ensure that all three operations (AND, OR, and ADD) are correctly computed using this value of gate delay.

Step 3. Construct a VHDL structural model of a 2-bit ALU. Use the single-bit ALU as a building block. Use a ripple carry implementation to propagate the carry between single-bit ALUs. Remember to analyze (compile) the single-bit model before you analyze (compile) the 2-bit model.

Step 4. Use the **generic map** construct in the structural model of the 2-bit ALU to set the value of the gate delay for the OR gates to 4 ns and the AND gates to 2 ns.

Step 5. Compile, simulate, and verify the functionality of this model.

Step 6. Open a trace window with the signals you would like to trace. In this case, you will need to only trace the input and output signals to test the model.

Step 7. Note how easy it is to modify the values of the gate delay at the top level. Now change the model to use a generic n-input AND gate. The model can be modified as follows:

> **Step 7 (a)** Modify the model of the AND gate to follow the model shown in Figure 5-12.

> **Step 7 (b)** Modify the single-bit ALU model to include a generic parameter specifying a default value of 3 for the number of inputs to the AND gate.

> **Step 7 (c)** In the model of the single-bit ALU declare the generic AND model as a component and replace the CSA describing the operation of the AND gate with a component instantiation statement for the generic AND model. This instantiation statement should also include a **generic map** statement providing the number of gate inputs as a parameter.

> **Step 7 (d)** Since the default value of the number of inputs to the AND gate is 3, you must pass parameters correctly for this model to function.

Step 8. Compile and test your model. Trace the input and output signals to determine that the model functions correctly.

Step 9. Experiment with other possibilities. For example, you can use a generic model of an OR gate as well. This model is shown in Figure 5-12.

End Simulation Exercise 5.3

5.5 Configurations

We have seen that there are many different ways in which to model the operation of a digital circuit utilizing behavioral and structural models. Approaches to the construction of behavioral models may differ in the use of concurrent and/or sequential statements. Structural models may employ multiple levels of abstraction and each component within a structural model may, in turn, be described as a behavioral or structural model. Consider the structural model of the full adder shown in Figure 5-2. Assume that two alternate architectures exist for the half-adder components. Design–1 is a behavioral model, as specified in Figure 3-2. Design–2 is a structural model, as specified in Figure 5-6. When the full-adder model shown in Figure 5-2 is compiled and simulated, which architecture for the half adder should be used? You would like to configure your simulation model to be able to use one or the other. The VHDL language provides *configurations* for explicitly associating an architecture description with each component in a structural model. This process of association is referred to as *binding* an instance of the component (in this example, the half adder) to an architecture. In the absence of any programmer-supplied configuration information, *default* binding rules apply.

Example: Component Binding

As an example of binding architectures to components, consider the structural model of a state machine for bit-serial addition, shown in Figure 5-4. The component C1 implements the combinational logic portion of the state machine. There may be alternative implementations of the gate-level design of C1 for high speed, low power, parts from different vendors, or even simply a behavioral model written for simulation. One of these alternative models must be *bound* to the component C1 for simulation. The configuration construct in VHDL specifies one, and only one, such binding.

   Note that we are concerned only with binding the combinational logic component with an architecture and are not concerned with configuring the entity description of C1. This is because the interface does not change. It is the implementation of C1 that may change and this is captured in an architecture for C1.

Example End: Component Binding

   Notice how easy it is to analyze different implementations. We simply change the configuration, compile, and simulate. Configurations also make it easy to share designs. When newer component models become available we can bind the new architecture to the component by editing the configuration and compile and simulate.

   There are two ways in which configuration information can be provided: *configuration specification* and *configuration declaration*. But first, let us state the default binding rules that have been in effect for the examples in this chapter and explain how the VHDL tools find the architectures for the components in a structural model when no configuration information is provided.

5.5.1 Default Binding Rules

The structural model is simply a description of a schematic. Revisiting our analogy with the construction of a circuit on a protoboard, we now have to build the operational circuit. Assume that each component can be realized with a single chip from one of many vendors. We can think of the configurations as describing the chips we need to obtain and

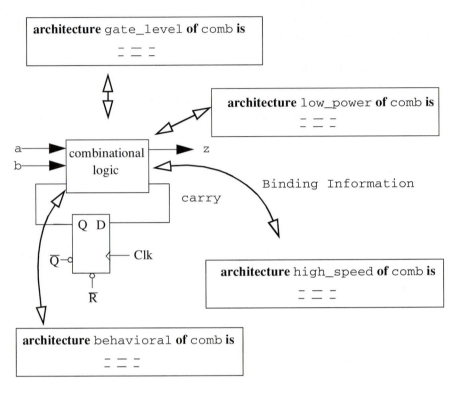

FIGURE 5-15 Alternative architectures for binding the combinational logic
component of a state machine

place on the board: one for each component. What if no configuration information was
provided? How would we know what chip to use for each component and where to get
them from? Clearly, we expect some rules for doing so. For example, we might first look
around the lab bench for chips with the same names as the components. If we could not
find them we might then look in the "usual" cabinet where all chips are stored, again looking for chips with the same names as the component names. Such rules are analogous to
the default binding rules in VHDL.

If no configuration information is provided as in the preceding examples then a
default architecture may be found as follows: if the entity name is the same as the component name, then this entity is bound to the component. For example, for the structural
model in Figure 5-4, no configuration information is provided. The language rules enable
a search for entities with the same names, in this case, comb and dff. These may be
found in the working directory, in which case the architectures associated with these entities are used. However, what if there are multiple architectures for the entity, such as for
comb as shown in Figure 5-15? In this case, the last compiled architecture for entity comb
is used. Now the environment can construct a complete simulation model. Using the

default rules, we see that by using component names that are the same as the entity names we can avoid writing configurations. This is what we have managed to do up to this point in this chapter. Using the same names for components and their entities also improves the readability of the code.

However, if no such entity with the same name is visible to the VHDL environment, then the binding is said to be deferred, that is, no binding takes place now, but information will be forthcoming later, as described in the following sections. This is akin to going ahead with wiring the rest of the circuit and hoping your partner comes up with the right chips before you are ready to run the experiment!

5.5.2 Configuration Specification

Configuration specifications are used in the architecture body to identify the relationships between the components and the entity–architecture pairs to be used to model each component. Continuing with our laboratory analogy, consider how we might specify the chips to be used for a component that we have declared. We might specify the chip name and location, for example, in the box labeled half adders in the grey cabinet. How can we similarly define the exact location of an entity–architecture pair? We can do so by naming the design library within which it is located and the name of the design unit within which they are stored. The syntax is shown in Figure 5-16.

For the first half-adder component we state that the entity description can be found in the library WORK. Libraries are generally implemented as directories in most systems. The library WORK is a special library and is usually the default working directory. Since an entity can have multiple architectures, the name of the architecture to be used for this entity can be specified within parentheses. Notice that for the second half adder, the configuration specification uses a different architecture. If no architecture is specified, then the last compiled architecture for that entity is used. All components of the same type need not use the same architecture, or even the same entity! The configuration of the 2-input OR gate deserves special attention. In this case, we have chosen to use an entity with a name different from the component. Therefore we have to provide additional information. For example, if you were simply given a new chip named lpo2, you would need to know which pins corresponded to inputs and which corresponded to outputs before you could use this in place of the or_2 component. This information is provided as part of the configuration statement, as shown in the figure. We are using an entity named lpo2 that has been compiled into a library named POWER. The corresponding architecture that we will use is named behavioral and can be found in the same library. Similar arguments hold for any generic parameters.

Thus, we see that we can specify any type of binding as long as the ports and arguments match up. We do not even need to use the same names. However, just as wiring up a circuit becomes a bit more involved, so does the writing and management of the models. The readability of the code is particularly important as the sizes of the models grow. The choice of names does become important. If we keep the same names for the entities and the components that are bound to them, then we can rely on default rules for configuring the simulation models and no **port map** clause is required in the configuration statement.

```
library IEEE;
library POWER;   -- a new library
use IEEE.std_logic_1164.all;
entity full_adder is
port (In1, In2, c_in : in std_logic;
      sum, c_out : out std_logic);
end full_adder;

architecture structural of full_adder is
component half_adder
port (a, b : in std_logic;
     sum, carry : out std_logic);
end component;

component or_2
generic (gate_delay : Time:= 2 ns);
 port (a, b : in std_logic;
      c : out std_logic);
end component;
signal s1, s2, s3 : std_logic;
--
-- configuration specification
--
for H1: half_adder use entity WORK.half_adder (behavioral);
for H2: half_adder use entity WORK.half_adder (structural);
for O1: or_2 use entity POWER.lpo2 (behavioral)
generic map(gate_delay => gate_delay)
port map (I1 => a, I2 => b, Z=>c);

begin         -- component instantiation statements
H1: half_adder port map (a =>In1, b => In2,
                              sum => s1, carry=> s2);
H2:  half_adder port map (a => s1, b => c_in,
                              sum => sum, carry => s2);
O1:  or_2 port map(a => s2, b => s3, c => c_out);

end structural;
```

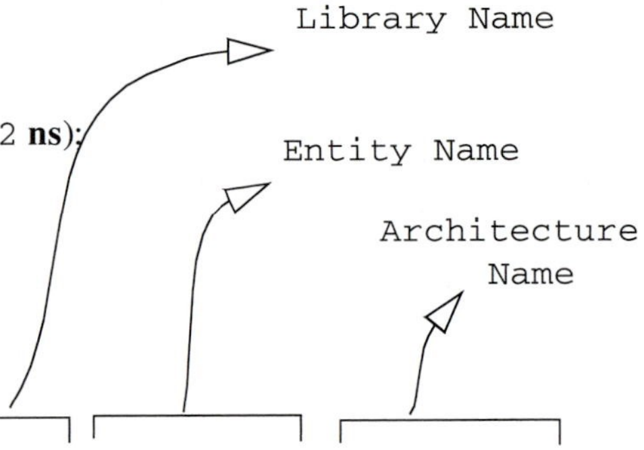

Library Name

Entity Name

Architecture Name

FIGURE 5-16 An example of using configuration specifications for the structural model of a full adder

```
configuration Config_A of full_adder is  -- name the configuration
                                          -- for the entity
for structural   -- name of the architecture being configured
for H1: half_adder use entity WORK.half_adder (behavioral);
end for;
--
for H2: half_adder use entity WORK.half_adder (structural);
end for;
--
for O1: or_2 use entity POWER.lpo2 (behavioral)
generic map(gate_delay => gate_delay)
port map (I1 => a, I2 => b, Z=>c);
end for;
--
end for;
end Config_A;
```

FIGURE 5-17 A configuration declaration for the structural model of the full adder in Figure 5-16.

5.5.3 Configuration Declaration

The configuration specification is part of the architecture and must be placed within the architecture body. Modification of our choice of models to implement a component requires editing the architecture and recompiling the model. A configuration declaration enables us to provide the same configuration information, but as a separate design unit. In the same way that entities and architectures are design units, so are configuration declarations. Suppose we take all of the configuration information provided in the architecture in Figure 5-16, label it, and refer to it by its label. This unit is a configuration declaration and is a distinct design unit. An example of the configuration information in Figure 5-16 provided as a configuration declaration, is shown in Figure 5-17.

Like other design units, we name configuration declarations as shown on the first line. In addition, the entity that is to utilize this configuration information is also named. Note that this declaration looks very similar to an architecture declaration. The second line identifies the name of the specific architecture of this entity that is being configured. For example, there could be another structural model of the full adder placed in an architecture labeled Structural_B. We must be able to distinguish between alternative architectures unambiguously and do so by referring to the unique architecture labels.

A close examination of the syntax will reveal that the **for** statements are terminated by **end for** clauses. While the above declaration deals only with one level of the hierarchy, configuration declarations can be written to span a complete design hierarchy with nested **for...end for** constructs to bind components at all levels of the hierarchy. It is also apparent that we can have different configurations for the full adder. For example, we might have a Config_B and a Config_C. Each configuration would use a set of components or models for the half adder and 2-input OR gate components. We might be motivated to take this approach to study the implementation with different technologies, for example, low-power vs. high-speed implementations.

There are several other advanced topics in the area of binding components to architectures, such as direct instantiation, incremental binding, and binding to configurations rather than an entity–architecture pair. These topics can be found in any advanced text on VHDL.

Simulation Exercise 5.4: Use of Configurations

This exercise will emphasize the need and importance of configurations. The exercise builds on Simulation Exercise 5.2 which produced two distinct models of an 8-bit adder. The first model was a structural model hierarchically built from smaller size ALUs. The second was a behavioral model constructed using processes.

Step 1.   Construct a 16-bit ALU from two 8-bit ALUs using ripple carry. Use the entity description of the 8-bit ALU developed in Simulation Exercise 5.2.

Step 2.   Include in the model a configuration specification such as the one shown in Figure 5-16, to specify the name of the architecture of the 8-bit model that you wish to use. Start with the hierarchical 8-bit model. All of the models that are used in building this 8-bit ALU are assumed to have been compiled into your working directory. Make sure the library WORK is set to your current working directory.

Step 3.   Compile the 16-bit ALU. Test the model and ensure that it is functioning correctly.

Step 4.   Now modify the configuration specification to use the behavioral model of the 8-bit ALU. This should require editing one line (in fact one word) of your VHDL model—the line in your configuration specification.

Step 5.   Test this model and ensure that it is working. Note the ease with which it is possible to "plug" in different models of subcomponents of the 16-bit ALU.

Step 6.   List some of the differences between the two models that you have constructed.

End Simulation Exercise 5.4

5.6 Common Programming Errors

The following are some common programming errors.

- Modifying the model of a component and forgetting to reanalyze the component model prior to reuse.

- Generics can have their values defined at three places: within the model, in a component instantiation statement using the **generic map**() construct, and within an architecture in a component declaration. Changing the value of the generics in one place may not have the intended effect due to precedence of the other declarations. The actual value of the generic parameter may not be what you expect.

- When using default bindings of components, the name, type, and mode of each signal in the component declaration must exactly match that of the entity, otherwise an error will result.

- Inheriting a generic value by way of default intializations in the component declarations in the higher level may lead to unexpected values of the generic parameters. A clear idea of how generic values are propagated through the hierarchy is necessary to ensure that the acquired values of the generic parameters corresponds to the intended values.

5.7 Chapter Summary

The focus of this chapter has been on the ability to specify hierarchical models of digital systems *ignoring* how the internal behavior of components may be specified. Internal behavior can be described using language features described in Chapter 3 and Chapter 4. An important aspect of the construction of hierarchical models is the ability to construct parameterized behavioral models and to be able to determine values of parameters by passing information down this hierarchy. This is facilitated by the **generic** construct and enables the construction of libraries of models that can be shared by designers. Finally, given such a library of alternative models for a component, the specification of the particular models to be used in the construction of a model is provided by **configuration** construct.

The concepts covered in this chapter include

- Structural models
  - component declaration
  - component instantiation
- Construction of hierarchical models
  - abstraction
  - trade-offs between accuracy and simulation speed

- Generics
    - specifying generic values
    - constructing parameterized models
- Configurations
    - component binding
    - default binding rules
    - configuration specification
    - configuration declaration

We now have command of the basic constructs for creating VHDL models of digital systems. The following chapters address remaining issues in support of these basic constructs to provide a complete set of modeling tools.

Exercises

1. Complete the structural model of the bit-serial adder shown in Figure 5-4 by constructing a model for the two components. You must complete the design of the combinational logic component. Compile, simulate, and test the model.

2. Consider the detailed hierarchical model of an 8-bit ALU constructed in Simulation Exercise 5.2. Now consider a single-level, behavioral model of the 8-bit ALU constructed using a single process and 8-bit data types. Compare the two models and comment on the simulation accuracy, simulation time, and functionality.

3. You are part of a software group developing algorithms for processing speech signals for a new digital signal processing chip. To test your software your options are to construct i) a detailed hierarchical model of the chip comprised of gate level models at the lowest level of the hierarchy or ii) a behavioral level model of the chip that can implement the algorithms that you wish to use. Your goal is to produce correct code for a number of algorithms prior to detailed testing on a hardware prototype. How would you evaluate these choices and what are the trade-offs in picking one approach over the other?

4. Construct and test a structural model of the circuit shown below. Note that there are many different ways in which to do this. You might consider each gate a component or groups of gates as a component represented by the Boolean function which is computed by this gates.

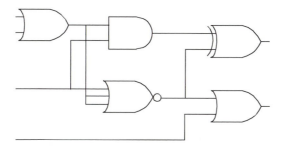

5. Consider the circuit shown below. Construct a structural model comprised of two components: a generic N-input AND gate and a two-input OR gate. By passing the appropriate generic value we can instantiate the same basic AND gate component as a two-input or three-input AND gate.

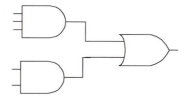

6. In problem 5 use generics to set default gate delays for the components. Now instantiate each AND gate with different gate delays using the **generic map** construct.

7. Modify the generic model of an N-bit register shown in Figure 5-13 to operate as counter that is initialized to a preset value.

8. Compare the use of configuration specifications and configuration declarations. When is one or the other advantageous?

9. Implement the structural model of a full adder using configuration specifications as shown in Figure 5-16. Use a simple model of the two-input OR gate rather than the one shown in the figure. You can omit the configuration statement for the OR gate and use the default binding for this component. Use two different architectures for the half-adder components.

CHAPTER 6 — Subprograms, Packages, and Libraries

With any large body of software we need mechanisms for structuring programs, reusing software modules, and otherwise managing design complexity. In conventional languages, mechanisms for doing so have been available to us for some time. The VHDL language also provides support for such mechanisms through the definition and use of procedures and functions for encapsulating commonly used operations. However, the presence of the signal class of objects, the fact that the programs represent the discrete event simulation of physical systems, and the notion of simulation time generate considerations that do not arise for their counterparts in conventional programming languages. For example, where are procedures declared and used? Can wait statements be used in a procedure? How are signals passed and modified? Can functions operate on signals? The essential issues governing the use of functions and procedures are initially discussed in this chapter.

Related groups of functions and procedures can be aggregated into a module that can be shared across many different VHDL models. Such a module is referred to as a *package*. In addition to the definitions of procedures and functions, packages may contain user-defined data types and constants and can be placed in *libraries*. Libraries are repositories for design units in general, and packages are a type of design unit. Collectively, procedures, functions, packages, and libraries provide facilities for creating and maintaining modular and reusable VHDL programs.

```
function rising_edge (signal clock: std_logic) return boolean is
--
--declarative region: declare variables local to the function
--
begin
-- body
--
return (expression)
end rising_edge;
```

FIGURE 6-1 Structure of a function

6.1 Essentials of Functions

As in traditional programming languages, functions are used to compute a value based on the values of the input parameters. An example of a function declaration is:

function rising_edge (**signal** clock: **in** std_logic) **return boolean**;

The function definition provides a function name, specification of the input parameters and the type of the result. Functions return values that are computed using the input parameters. Therefore, we would expect that the parameter values are used, but not changed within the function. This notion is captured in the **mode** of the parameter. Parameters of mode **in** can only be read. Functions cannot modify parameter values (procedures can) and therefore functions do not have any parameters of mode **out**. Since the mode of all function parameters is **in**, we do not have to specify the mode of a parameter. A discussion of other modes is presented with a discussion of procedures in Section 6.2.

Consider the structure of a function as shown in Figure 6-1. The function has a name (rising_edge) and a set of parameters. The parameters in the function definition are referred to as *formal* parameters. Formal parameters can be thought of as placeholders that describe the type of object that will be passed into the function. When the function is actually called in a VHDL module, the arguments in the call are referred to as *actual* parameters. For example, the above function may be called in the following manner:

```
rising_edge (enable);
```

In this case, the actual parameter is the signal enable, and takes the place of the formal parameter clock in the body of the function. The type of the formal and actual parameters must match—except for formal parameters which are constants. In this case, the actual parameter may be a variable, signal, constant, or an expression. When no class is specified, the default class of the parameter is constant. Wait statements are not permitted in functions. Thus, functions execute in zero simulation time. It follows that wait statements cannot exist in any procedures called by a function (although procedures are allowed to have wait statements). Furthermore, parameters are restricted to be of mode **in**, and therefore functions cannot modify the input parameters. Thus, signals passed into functions cannot be assigned values. This behavior is consistent with the conventional definition of functions.

Example: Detection of Signal Events

Often we find it useful to perform simple tests on signals to determine if certain events have taken place. For example, the detection of a rising edge is common in the modeling of sequential circuits. Figure 6-2 shows the VHDL model of a positive edge triggered D flip-flop from Figure 4-7. The only difference is the inclusion of a function for testing for the rising edge, rather than having the function code in the body of the VHDL description. Note the placement of the function in the declarative portion of the architecture. Normally this region is used to declare signals and constants used in the body of the code. Therefore, we might expect that we can also declare functions (or procedures) that are used in the architecture too. This is indeed the case. The function could have also been declared in the declarative region of the process that called the function (i.e. between the keywords **process** and **begin**.) The question is whether you wish to have the function be visible to, and therefore callable from, all processes in the architecture body, or visible to just one process. In practice, we would much rather place related functions and procedures in packages: a type of design unit described later in this chapter. In fact, such a function is provided in the package `std_logic_1164`.

Example End: Detection of Signal Events

6.1.1 Type Conversion Functions

Type conversion is another common instance of the use of functions. The model of the memory module in Section 4.1 represented memory as a one-dimensional array. This array is indexed by an integer. However, memory addresses are provided as an n-bit binary address. We find that we need to convert this bit vector representing the memory address to an integer used to index this array representing memory. In other instances we may want to use models of components developed by others. However, let us say that they have used signals of type **bit** and **bit_vector** while you have been using signals of type `std_logic` and `std_logic_vector`. If it is possible (and correct) to use their models, type conversion functions will be necessary for interoperability if we do not wish to invest in the time to convert their models to use the IEEE 1164 types.

```
library IEEE;
use IEEE.std_logic_1164.all;
entity dff is
port (D, Clk : in std_logic;
      Q, Qbar : out std_logic);
end dff;

architecture behavioral of dff is
function rising_edge (signal clock : std_logic) return boolean is
variable edge : boolean:= FALSE;
begin
edge := (clock = '1' and clock'event);
return (edge);
end rising_edge;

begin
output: process
begin
wait until (rising_edge(Clk));

    Q <= D after 5 ns;
    Qbar <= not D after 5 ns;

end process output;
end behavioral;
```

FIGURE 6-2 An example of the use of functions

Example: Type Conversion

Consider the VHDL type **bit_vector** and the IEEE 1164 type std_logic_vector. We may wish to make assignments from a variable of one type to a variable of the other type. For example, consider the VHDL model of a memory system shown in Figure 4-2. In this model we use the function to_stdlogicvector() that is provided in the package std_logic_1164.vhd (more on packages in Section 6.4). This function takes as an argument an object of type **bit_vector** and returns a value of type std_logic_vector. Conversely we may wish to convert from **bit_vector** to std_logic_vector. An example of the implementation of this function is shown in Figure 6-3. The function simply scans the vector and converts each element of the input std_logic_vector. The type **bit** may take on values 0 and 1, whereas the type std_logic may take on one of nine

```
function to_bitvector (svalue : std_logic_vector) return
bit_vector is
variable outvalue : bit_vector (svalue'length-1 downto 0);
begin
for i in svalue'range loop -- scan all elements of the array
case svalue (i) is
when '0' => outvalue (i) := '0';
when '1' => outvalue (i) := '1';
when others => outvalue (i) := '0';
end case;
end loop;
return outvalue;
end to_bitvector
```

FIGURE 6-3 An example of a type conversion function

values. Note the declaration of the variable `outvalue`. The function declaration does not provide the size of the number of bits in the argument. This is set when the formal parameter is associated with the actual parameter at the time the function is called. In this case, how can we declare the size of any local variable that is to have the same number of bits as the input parameter? The answer is: by using attributes. As discussed in Section 4.5, arrays have an attribute named **length**. The value of `svalue'`**length** is the length of the array. By using unconstrained arrays in the definition of the function, we can realize a flexible function implementation where the actual size of the parameters are determined when the actual parameters are bound to formal parameters, which occurs when the function is called.

There are many ways in which to perform such conversions, and the figure shows but one of them. By examining commercial packages such as `std_logic_arith`.vhd and `std_logic_1164`.vhd we will find many such conversion functions. Examples include conversion from `std_logic_vector` to **integer** and vice-versa. Check the libraries that come with the installation of your VHDL simulator and tools. You will find many packages and it is useful to browse through them and study the procedures and functions that are contained within them.

Example End: Type Conversion

6.1.2 Resolution Functions

Resolution functions comprise a special class of functions. Recall from Chapter 4 that the resolved type is a signal that may have multiple drivers. This occurs quite often in digital systems. For example, consider a high-level model of a computer system shown in Figure

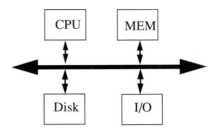

FIGURE 6-4 A simple model of a computer

6-4. The CPU, memory, and some peripherals such as a disk and other I/O devices must exchange data. It is quite expensive to have a dedicated interconnect between every pair of communicating devices. Furthermore, all of the devices do not necessarily have to communicate at the same time. A common architecture is to have these devices communicate over a shared set of signals called a bus. In the VHDL model of such an architecture, the shared bus could be a signal datatype and several components may schedule values on this bus during the course of a simulation. In other words, the bus has multiple drivers. In order for the simulation to be correct, we must consider the possibility that models for various components on the bus may have scheduled events on the bus for the same time. What, then, is the value assigned to the signal at that time? How would an actual implementation perform?

Circuits that implement wired logic are another example of instances where a signal can have multiple drivers. The design is such that the signal value represents a logical operation such as a Boolean AND. To correctly simulate such circuits we must be able to unambiguously state the value of the signal at any point in time. The value must be *resolved* based on the values scheduled by the multiple drivers of the signal. The algorithm for resolving the issue of the signal value at any time is captured in the *resolution function*. One can think of the resolution function as examining all of the scheduled values on the bus for that time and determining the value of the signal. For example, the value could be a logical OR of all of the signals or the maximum value. The signal must be declared as a *resolved type*, which means that there is a resolution function associated with all signals of this type. This function should accurately reflect the behavior of the physical system being modeled. During the course of a simulation when any signal of this type is to be assigned a value, the resolution function is invoked. This function examines the values on all drivers for that signal and computes the correct signal value as defined by the resolution function. The resolution function must be an associative operation so that the order in which the multiple signal drivers values are examined does not affect the resolved value of the signal. For example, a logical OR operation on a set of values of type **bit** is an associative operation. Throughout this text we have used the signal type std_logic: a resolved type defined by the IEEE 1164 standard. The following example from the implementation of

```
type std_ulogic is ('U', -- Uninitialized
                     'X', -- Forcing Unknown
                     '0', -- Forcing 0
                     '1', -- Forcing 1
                     'Z', -- High Impedance
                     'W', -- Weak Unknown
                     'L', -- Weak 0
                     'H', -- Weak 1
                     '-' -- Don't care
          );
```

function resolved (s : std_ulogic_vector) **return** std_ulogic;

subtype std_logic **is** resolved std_ulogic;

FIGURE 6-5 An example of the declaration of resolved signals

the IEEE 1164 standard illustrates how resolved types can be declared and how resolution functions can be defined.

Example: Resolved Types in the IEEE 1164 Standard

Let us examine the definition and use of the resolved type std_logic from an implementation of the IEEE 1164 standard. Figure 6-5 shows an example of the declaration and use of resolved types taken from an implementation of the IEEE 1164 standard, std_logic_1164.vhd, that is provided with just about any VHDL toolset. This implementation of the standard first defines a new type: std_ulogic. A signal of this type takes on nine values, as defined in Figure 6-5 and as described in Section 2.4. This type is referred to as an enumerated type, since the list of values of an object of this type are explicitly enumerated. However, any signal declared to be of this type can support only a single driver. Therefore, we wish to define a new signal type that can take on all of the values of std_ulogic, but can also support multiple drivers. This is done on the following line by creating the (sub) type std_logic.

Consider the structure of this declaration. It looks very much like any other declaration except for two items. A *subtype* simply means that the declared signal can take on a range of values that is a subrange of the original or *base* type. Secondly, the type provides the name of a *resolution function* that is associated with all objects declared to be of this type. In this case it is the function named resolved. What this definition means is the following: whenever a signal of type std_logic is to be assigned a value, there may be multiple drivers associated with this signal. The values from these multiple drivers are passed to the resolution function, which determines the value to be assigned to the signal. For example, consider the case where a single-bit bus is being driven to a logic 1 by one driver left in a high-impedance state or Z by another driver. The value of the signal should be a logic 1. The resolution function must be capable of making this determination and of handling more than two drivers.

One simple approach to implementing a resolution function is to build a table. The row and column indices correspond to the signal values from two drivers. The table entry corresponds to the value that would be produced if these two drivers were attempting to drive a signal to these two values. For example, the entry in a table at location (Z,1) would be 1. Now, if we had a set of drivers, we could compute the final value by resolving the values of all drivers in a pairwise manner. The structure of the table used by an implementation of the IEEE 1164 standard to resolve the values of a pair signals of type std_logic is shown in Table 6-1. For example, two driver values of Z and W will yield a signal value of W.

TABLE 6-1 Table for resolving the values of a pair of signals of type std_logic

	U	X	0	1	Z	W	L	H	–
U	U	U	U	U	U	U	U	U	U
X	U	X	X	X	X	X	X	X	X
0	U	X	0	X	0	0	0	0	X
1	U	X	X	1	1	1	1	1	X
Z	U	X	0	1	Z	W	L	H	X
W	U	X	0	1	W	W	W	W	X
L	U	X	0	1	L	W	L	W	X
H	U	X	0	1	H	W	W	H	X
–	U	X	X	X	X	X	X	X	X

Example End: Resolved Types in the IEEE 1164 Standard

Example: Using Resolution Functions

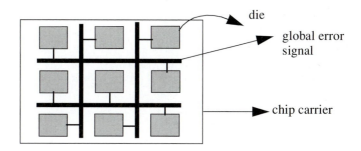

FIGURE 6-6 Structure of a multichip module with multiple die

Let us consider another example of the implementation of resolution functions. Consider a system with multichip modules (MCM): these are chip carriers that can support multiple semiconductor die on a single substrate and within a single package. Suppose that we have multiple die placed on a single multichip module, as shown in Figure 6-6. Assume that for this MCM module to be functional all of the die must be functional. This means that when one die has failed, the module is considered defective. Periodically self-test circuitry in the die (or even in the MCM substrate!) will locally run diagnostics to test the individual die to ensure that they remain functional. There is a single global error signal that is driven by an output of the self-test circuitry within each die. This global signal performs the logical OR of all of the single-bit diagnostic results from the individual die. If this global error signal is asserted, then at least one bad die exists on the MCM and the complete package is considered to be faulty. Since there are multiple drivers, this global signal must be of a resolved type. An example of the VHDL code that may be used to implement this signal type is shown in Figure 6-7.

In this model, we assume that the behavior of each chip is captured with a process. Note that we could have used a structural model instead. Each of the processes may place a value of 0 or 1 on the shared signal `error_bus`. If at least one driver forces `error_bus` to 1, we would like the value of `error_bus` to remain 1. This behavior is captured in the resolution function shown in Figure 6-7.

The function `wire_or` receives as input an array of values. The loop simply scans this array looking for the first 1. One very interesting feature about this function is that the parameter is an unconstrained array. This means that we have not specified the number of elements that will be passed to the function. Can we pass an array of 32 elements? 64 elements? The answer is yes to both. The size of the input parameter will be determined at the time the function is called. In this case we do not know how many signal drivers may exist. The use of unconstrained arrays in this manner is quite common. This body of the function simplifies the process of resolving signal values. The base type of `wire_or_logic` is

```
library IEEE;
use IEEE.std_logic_1164.all;

entity mcm is
end mcm;

architecture behavioral of mcm is
function wire_or (sbus :std_ulogic_vector) return std_ulogic;
begin
for i in sbus'range loop
 if sbus(i) = '1' then
return '1';
end if;
end loop;
return '0';
end wire_or;

subtype wire_or_logic is wire_or std_ulogic;
signal error_bus : wire_or_logic;
begin
Chip1: process
begin
-- ..
error_bus <= '1' after 2 ns;
-- ..
end process Chip1;
Chip2: process
begin
-- ..
error_bus <= '0' after 2 ns;
-- ..
end process Chip2;
end behavioral;
```

There could be many more processes like this, e.g., corresponding to each die

FIGURE 6-7 An example of the use of resolution functions

std_ulogic. Thus, any signal of type wire_or_logic can actually take on values other than 1 or 0. However, the resolution function shown here reflects a classical 0/1 view of single-bit signals and ignores these other values. A more realistic and robust approach would account for all possible combination of signal values.

Example End: Using Resolution Functions

The second class of subprograms are procedures. A distinguishing feature of procedures is that they can modify input parameters. In this case we must consider how signals are passed and handled. The essential issues in getting started with writing and using procedures are addressed in the next section.

6.2 Essentials of Procedures

Procedures are subprograms that can modify one or more of the input parameters. The following procedure declaration illustrates the procedure interface.

> **procedure** read_v1d (**variable** f: **in text**; v : **out** std_logic_vector);

This is a procedure to read data from a file where f is a file parameter. The first characteristic we might notice is that parameters may be of mode **out**. Just as parameters of mode **in** must be read and cannot be written, parameters of mode **out** cannot be read and used in a procedure but can only be written. We may also have parameters of mode **inout**. As with functions, the type of the formal parameters in a procedure declaration must match the type of the actual parameters that are used when the procedure is called. If the class of the procedure parameter is not explicitly declared, then parameters of mode **in** are assumed to be of class **constant**, while the parameters of mode **out** or **inout** are assumed to be of class **variable**. Variables declared within a procedure are initialized on each call to the procedure and their values do not persist across invocations of the procedure.

Example: Interface to Memory

Let us consider a VHDL model for a simple processor where we have two components: a CPU and memory. The behavioral model of the CPU must be able to read and write locations from memory. These operations are common candidates for implementation as procedures. We will create two procedures: one to read and one to write memory locations. We will assume the memory model and associated signals as shown in Figure 4-1, with the addition of one additional signal from memory that signifies the completion of a memory operation. These procedures are shown in Figure 6-8. Both procedures should manipulate the signals shown in the memory interface in Figure 4-1. There are a number of interesting features shown here. Note the presence of wait statements within the procedure. Thus, a process can suspend inside a procedure. Furthermore, signals can be assigned values within a procedure. This raises the issue of how signals are passed into a procedure, and this issue is dealt with in Section 6.2.1. Signals that are modified within the procedure are declared to be of mode **out**. For example, see signal R in procedure mread() in Figure 6-8.

The body of the architecture description is likely to include processes within which the procedure calls can be made. Alternatively, the procedures could have been declared

```vhdl
library IEEE;
use IEEE.std_logic_1164.all;
entity CPU is
port (DI : out std_logic_vector ( 31 downto 0);
    ADDR :out std_logic_vector (2 downto 0);
    R, W : out std_logic;
    DO : in std_logic_vector(31 downto 0);
    S : in std_logic);
end CPU;

architecture behavioral of CPU is
procedure mread (address : in std_logic_vector (2 downto 0);
            signal R : out std_logic;
            signal S : in std_logic;
            signal ADDR : out std_logic_vector (2 downto 0);
            signal data : out std_logic_vector (31 downto 0)) is
begin
ADDR <= address;
R <= '1';
wait until S = '1';
data <= DO;
R <= '0';
end mread;

procedure mwrite (address : in std_logic_vector (2 downto 0);
                signal data : in std_logic_vector (31 downto 0);
                signal ADDR : out std_logic_vector (2 downto 0);
                signal W : out std_logic;
                signal DI : out std_logic_vector (31 downto 0)) is
begin
ADDR <= address;
W <= '1';
wait until S = '1';
DI <= data;
W <= '0';
end mwrite;
--
-- any signal declarations for the architecture here
--
```

FIGURE 6-8 An example of the use of procedures

```
begin
--
-- CPU behavioral description here
process
begin
--
-- behavioral description
--
end process;

process
begin
--
-- behavioral description
--
end process;
end behavioral;
```

FIGURE 6-8 (cont.)

within the declarative region of the process: just before the **begin** statement and after the **process** statement. Just as processes can declare and use variables that are local to the process, processes may also declare and use procedures within a process. However, in this case they would be visible only within that process.

Example End: Interface to Memory

6.2.1 Using Procedures

Signals cannot be declared within procedures. However, signals can be passed into procedures as parameters. Due to visibility rules, procedures can make assignments to signals that are not explicitly declared in the parameter list. For example, procedures declared within a process can make assignments to signals corresponding to the ports of the encompassing entity. This is possible since the ports are visible to the process. The procedure is said to have side effects, since it has an effect on a signal that is not declared in the parameter list. This is poor programming practice, since it makes it difficult to reason about the models (e.g., when debugging) and understand their behavior. Clarity and understanding of the code is enhanced if parameters are passed explicitly rather than relying on side effects. If the class of a parameter is not declared, and the mode is **out** or **inout**, then the class defaults to that of a variable. If the mode is **in**, the class of the parameter defaults to a constant.

Procedures can also be placed in the declarative region of a process. We know that processes cannot have a sensitivity list and also have wait statements in the body of the process. Therefore, when we use procedures, it follows that a process that calls a procedure with a **wait** statement cannot have a sensitivity list.

6.2.2 Concurrent and Sequential Procedure Calls

Depending on how procedures are used, we can distinguish between concurrent and sequential procedure calls. Remember the concurrent signal assignment statements from Section 3.3.1? Each statement represented the assignment of a value to a signal, and this assignment occurred in simulated time concurrently with the execution of the other concurrent signal assignment statements and processes. Concurrent procedure calls can be viewed similarly. The procedure is invoked in the body of an architecture concurrently with other concurrent procedures, concurrent signal assignment statements, or processes. The procedure is invoked when there is an event on a signal that is an input parameter to the procedure. It follows that if we use a concurrent procedure, the parameter list cannot include a variable, since (in VHDL'87) variables cannot exist outside of a process. In contrast, sequential procedure calls are those where the procedure is invoked within the body of a process. In this case, the invocation of the procedure is determined by the sequence of execution of statements within the process—just like in a conventional program. The following example should help solidify our understanding of the differences between concurrent and sequential procedure calls.

Example: Concurrent and Sequential Procedure Calls

Figure 6-9 illustrates an example of a concurrent procedure call. The structural model of a bit-serial adder from Figure 5-4 has been rewritten such that the D flip-flop component instantiation statement has been replaced by a procedure. The procedure is invoked concurrently with the component comb whenever there are events on the signals that are declared to be of mode **in**. Thus, events on the clk, reset or d inputs will cause this procedure to be invoked. From the procedure body we see that the output is modified only on the rising edge of the clk signal.

This structure does appeal to our understanding of VHDL programs. Consider what would happen in the model in the example of Figure 5-4. When this model is simulated, let us assume that the component dff would be replaced by a behavioral model similar to the one shown in Figure 4-7. We see that the procedure effectively implements the same behavior. Note how the parameter list explicitly associates the formal and actual parameters rather than having this association made by virtue of the position in the call. The signal qbar is associated with the keyword **open** in the procedure call. This is akin to leaving a pin of a device (dff) unconnected.

Figure 6-10 shows the equivalent implementation as a sequential procedure call. The procedure is encased in a process with an explicit **wait** statement. Note the structure of the

```
library IEEE;
use IEEE.std_logic_1164.all;
entity serial_adder is
port (a, b, clk, reset : in std_logic;
      z : out std_logic);
end serial_adder;

architecture structural of serial_adder is
component comb
 port (a, b, c_in : in std_logic;
       z, carry : out std_logic);
end component;

procedure dff (signal d, clk, reset : in std_logic;
               signal q, qbar : out std_logic) is
begin
 if (reset = '0') then
    q <= '0' after 5 ns;
    qbar <= '1' after 5 ns;
   elsif (clk'event and clk = '1') then
   q <= d after 5 ns;
   qbar <= (not D) after 5 ns;
 end if;
end dff;

signal s1, s2 : std_logic;

begin
C1: comb port map (a => a, b => b, c_in => s1, z =>z, carry => s2);
--
-- concurrent procedure call
--
dff(clk => clk, reset =>reset, d=> s2, q=>s1, qbar =>open);
end structural;
```

FIGURE 6-9 An example of a concurrent procedure call

```
library IEEE;
use IEEE.std_logic_1164.all;
entity serial_adder is
port (a, b, clk, reset : in std_logic;
      z : out std_logic);
end serial_adder;

architecture structural of serial_adder is
component comb
 port (a, b, c_in : in std_logic;
        z, carry : out std_logic);
end component;
procedure dff (signal d, clk, reset : in std_logic;
                  signal q, qbar : out std_logic) is
begin
 if (reset = '0') then
    q <= '0' after 5 ns;
    qbar <= '1' after 5 ns;
   elsif (clk'event and clk = '1') then
   q <= d after 5 ns;
   qbar <= (not D) after 5 ns;
 end if;
end dff;
signal s1, s2 : std_logic;
begin
C1: comb port map (a => a, b => b, c_in => s1, z =>z, carry =>  s2);
process
begin
 dff(clk => clk, reset =>reset, d=> s2, q=> s1, qbar => open);
wait on clk, reset, s2;
end process;
end structural;
```

FIGURE 6-10 An example of a sequential procedure call

wait statement. If an event occurs on any of the signals in the list, the process will be executed, which in this case will cause the procedure dff() to be called. This procedure call model is equivalent to the model shown in Figure 6-9.

Example End: Sequential and Concurrent Procedure Calls

6.3 Subprogram and Operator Overloading

A very useful feature of the VHDL language is the ability to *overload* the subprogram name. For example, there are several models and implementations of a D flip-flop. We saw a few examples in Section 4.4. Imagine we were to write behavioral models of sequential circuits that included D flip-flops. We might be using procedures such as the one shown in Figure 6-9 to model the behavior of a D flip-flop. If we wished to incorporate models that had asynchronous set and clear signals, we might write another procedure with a different name, say `asynch_dff()`. What if we wish to have procedures that would operate on signal arguments of type **bit_vector** rather than `std_logic_vector`? We would write distinct procedures to incorporate models with these types and behaviors. By accommodating various possibilities of argument types and flip-flop behavior we might have to write many different procedures while keeping track of the names to distinguish them.

It would be very helpful to be able to use a single name for all procedures describing the behavior of various types of D flip-flops. We would like to call `dff()` with the right parameters and let the compiler determine which procedure to use based on the number and type of arguments. For example, consider the two procedure calls

```
dff(clk, d, q, qbar)
```

```
dff(clk, d, q, qbar, reset, clear)
```

From the arguments, we can see that we are referring to two different procedures, one that utilizes asynchronous reset and clear inputs and one that does not (e.g., corresponding to Figure 4-8 and Figure 4-7 respectively). From the type and number of arguments we can tell which procedure we meant to use. This process is referred to as overloading subprogram names or simply *subprogram overloading*. When we create such a set of procedures or functions with overloaded names, we would probably place them in a package (see Section 6.4) and make the package contents visible via the **use** clause. If we examine the contents of some of the packages shown in Appendix D, we will see examples of overloaded functions and subprograms. For example, note that in `std_logic_1164.vhd` the Boolean functions **and**, **or**, etc. have been defined for the type `std_logic`.

Similarly, the operators such as "*", and "+" have been defined for certain predefined types of the language such as integers. What if we wish to perform such operations on other data types that we may create? We can overload these operators by providing definitions of "*" and "+" for these new data types. CAD tool vendors typically distribute packages that contain definitions of operators and subprograms for various operations on data types that are not predefined for the language. For example, the `std_logic_arith.vhd` package distributed by CAD tool vendors provides definitions for various operators over the `std_logic` and `std_logic_vector` types. Two examples of overloading the definitions of "*****" and "**+**" operators taken from this package are:

function "*****" (arg1, arg2: std_logic_vector) **return** std_logic_vector;
function "**+**" (arg1, arg2 :signed) **return** signed;

This means that if the contents of this package is included in a model via the **use** clause, then statements such as

s1 <= s1 + s2;

are valid, where all three signals are of type std_logic. Otherwise, the "**+**" operation is not defined for objects of type std_logic and this statement would be in error.

Procedures and functions are necessary constructs for building reusable blocks of VHDL code, for hiding design complexity, and for managing large complex designs. As we have seen in this section, they are also a means for enriching the language to easily handle new data types by encapsulating the definitions of common operations and operators over these data types. Even with a small number of new data types, the need to overload all of the common operators can generate quite a large number of functions. Furthermore, when we think of overloading subprogram names we can generate quite a few additional procedures and/or functions. Packages are a mechanism for structuring, organizing, and using such user-defined types and subprograms. These concepts are discussed next.

6.4 Essentials of Packages

As we acquire larger groups of functions and procedures within the models that we construct, we must consider how they will be used. We can use text editors and manually insert these functions into the VHDL models as we use them. This is a rather tedious process at best, especially when the models we construct grow large. A better approach would be to group logically related sets of functions and procedures into a module that can be easily shared among distinct designs and people. *Packages* are a means for doing so within the VHDL language.

Packages provide for the organization of type definitions, functions, and procedures so that they can be shared across distinct VHDL programs. To gain an intuition for the constructs used in building packages, it is instructive to consider how we try to reuse code modules across projects. When we are working on large class projects we attempt to make the most efficient use of our time by reusing functions or procedures that we may have written for older programs, have found somewhere on the Internet, or garnered from friends. For example, imagine that you have painstakingly put together a package that contains useful functions, procedures, and data types to help designers build simulation models of common computer architectures. This package may include definitions of new types for registers, instructions, and memories, as well as procedures for reading and/or writing memories, procedures for performing logical shift operations, and functions for type conversion operations. After months of tedious development you are now interested in promoting its use among fellow VHDL developers. How might you communicate the contents of this package in convincing them of its utility? What would developers want or need to know to determine if they could benefit from using the contents of your package? At the very least we would need to have a list of the functions and procedures and what they do. For example, for each procedure what values are computed and returned, and what parameters must be passed to perform these computations. This information forms the basis of the *package declaration*. It forms the interface or specification of the services that your package provides. When we write C or VHDL programs we must declare the variables or signals that we using: their type and possibly initial values. Similarly, when we write packages we must declare their contents. It is just that their contents are now more complex objects, such as functions, procedures, and data types. The package declaration is the means by which users declare what is available for use by VHDL programs. In the same sense that a hardware design unit possesses an external interface to communicate with other components, the package declaration defines the interface to other VHDL design units.

The easiest way to understand packages is by example, so let us examine a package that provides a new data type and a set of functions that operate on that data type. Throughout this text, the examples have declared and used the package `std_logic 1164.vhd`. Now let us look inside this package to see how the type is declared and how functions and declarations are used. Figure 6-11 shows a portion of the package declaration of an implementation of the IEEE 1164 package distributed with the vast majority of the VHDL environments. The listing of the package declaration is provided in Appendix D.3. We know that the basic VHDL type **bit** can take on only values 0 and 1 and therefore is inadequate to represent most real systems. By using the concept of enumerated types (see Chapter 9) a new type, `std_ulogic`, is defined as shown in Figure 6-11. Now a signal can be declared to be of this type:

signal *example_signal* : `std_ulogic`: = 'U';

The signal *example_signal* can now be assigned any one of the nine values defined above rather than the two values 0 and 1. The package also declares a resolved type, `std_logic`, which is a subtype of `std_ulogic`. This declaration simply states that in the course of the simulation when a signal of type `std_logic` is assigned a value, the resolution function `resolved` will be invoked to determine the correct value of a signal from the multiple drivers associated with the signal. However, we do have a problem in that all of the predefined logical functions, such as AND, OR, and XOR, operate on signals of type **bit,** which is predefined by the language. These logic functions must be redefined for signals of the above type. Some of these functions are shown in Figure 6-11. The declaration of all of the functions provided in this package can be found in Appendix D.3. If we are constructing a package that uses types, procedures, or functions from another package, then access to this package must be provided via **library** and **use** clauses.

Now that we have defined what is in the package, we must provide the VHDL code that implements these functions and procedures. This implementation is contained in the

```
package std_logic_1164 is
    ----------------------------------------------------------------
    -- logic state system (unresolved)
    ----------------------------------------------------------------
    type std_ulogic is ('U', -- Uninitialized
                        'X', -- Forcing Unknown
                        '0', -- Forcing 0
                        '1', -- Forcing 1
                        'Z', -- High Impedance
                        'W', -- Weak Unknown
                        'L', -- Weak 0
                        'H', -- Weak 1
                        '-'  -- Don't care
                    );
    type std_ulogic_vector is array (natural range <>) of
    std_ulogic;

    function resolved (s : std_ulogic_vector) return
    std_ulogic;
    subtype std_logic is resolved std_ulogic;

    type std_logic_vector is array (natural range <>) of
    std_logic;

    function "and" (l, r : std_logic_vector) return
    std_logic_vector;
    function "and" (l, r : std_ulogic_vector) return
    std_ulogic_vector;
    --
    --..<rest of the package definition>
    --
end std_logic_1164;
```

FIGURE 6-11 Examples from the package declaration of an implementation of the IEEE 1164 standard

package body. The package body is essentially a listing of the implementations. The body is structured as follows:

> **package body** my_package **is**
>
> --
>
> -- type definitions, functions, and procedures
>
> --
>
> **end** my_package;

Once we have these packages how do we use them? They are typically compiled and placed in *libraries* and referenced within VHDL design units via the **use** clause. All of the examples in this text have utilized the package std_logic_1164.vhd, which is in the library named IEEE. The essential properties of libraries are discussed next.

6.5 Essentials of Libraries

Each design unit—such as an entity, architecture, package body—is analyzed (compiled) and placed in a *design library.* Libraries are generally implemented as directories and are referenced by a logical name. In the implementation of the VHDL simulator, this logical name maps to a physical path to the corresponding directory and this mapping is maintained by the host implementation. However, just like variables and signals, before we can use a design library we must declare the library we are using by specifying the library's logical name. This is done using the library clause that has the following syntax:

> **library** *logical-library-name-1, logical-library-name-2,...;*

In VHDL, the libraries STD and WORK are implicitly declared. Therefore user programs do not need to declare these libraries. The former contains standard packages provided with VHDL distributions. The latter refers to the working directory which can be set within the simulation environment you are using. Refer to your simulator documentation on how this can be done. However, if a program were to access functions in a design unit that was stored in a library with the logical name IEEE, then this library must be declared at the start of the program. Most, if not all, vendors provide an implementation of the library IEEE with packages such as std_logic_1164.vhd, as well as other mathematics and miscellaneous packages.

Once a library has been declared all of the functions, procedures, and type declarations of a package in this library can be made accessible to a VHDL model through the **use** clause. For example, the following statements appear in all of the examples in this text:

> **library** IEEE;
>
> **use** IEEE.std_logic_1164.all;

The second statement makes *all* of the type definitions, functions, and procedures defined in the package std_logic_1164.vhd visible to the VHDL model. It is as if all of the declarations had been physically placed within the declarative part of a process that uses them.　A second form of the **use** clause can be used when only a specific item, such as a function called `my_func`, in the package is to be made visible.

　　　　use `IEEE.std_logic_1164.my_func;`

　　The **library** and **use** clauses establish the set of design units that are visible to the VHDL analyzer as it is trying to analyze and compile a specific VHDL design unit. When we first start writing VHDL programs we tend to think of single entity–architecture pairs when constructing models. We probably organize our files in the same fashion with one entity description and the associated architecture description in the same file. When this file is analyzed, the **library** and **use** clauses determine which libraries and packages within those libraries are candidates for finding functions, procedures, and user defined types that are referenced within the model being compiled. However, these clauses apply only to the immediate design unit. *Visibility must be established for each design unit separately!* Every design unit must be preceded by **library** and **use** clauses as necessary. If we start having multiple design units within the same physical file, then each design unit must be preceded by the **library** and **use** clauses necessary to establish the visibility to the required packages. For example, let us assume that the VHDL models shown in Figure 6-2 and Figure 6-7 are physically in the same file. The statements

　　　　library `IEEE;`

　　　　use `IEEE.std_logic_1164.all;`

　　must appear at the beginning of each model. We cannot assume that since we have these statements at the top of the file, they are valid for all design units in the same file. In this case if we neglected to precede each model with the preceding statements, the VHDL analyzer would return with an error on the use of the type `std_logic` in the subsequent models　since this is not a predefined type within the language, but rather is defined in the package `std_logic_1164.vhd`.

Simulation Exercise 6.1: Packages and Libraries

This exercise concerns creating and using a simple package.

Step 1.　Using a text editor create a package with the following characteristics.

　　　　Step 1 (a) Include several procedures for simulating a D flip-flop. You can start with the basic procedure given in Section 4.4 and modify this to produce procedures for the following

　　　　　　– arguments of type **bit** and `std_logic`

　　　　　　– arguments of type **bit_vector** and `std_logic_vector` (these are registers)

– use of reset and clear functions for different types of arguments

Step 2. Analyze and test each of the procedures separately before committing them to placement within the package.

Step 3. Define a new type designed to represent a 32-bit register. This is simply a 32-bit object of the type `std_logic_vector`.

type `register32` **is** `std_logic_vector` (31 **downto** 0);

Step 4. Propose and implement one or two other types of objects that you may expect to find in a model of a CPU or memory system.

Step 5. Create a library named `MYLIB`. This operation is usually simulator specific.

Step 6. Compile the package into the library `MYLIB`. This is typically done by setting the library `WORK` to be `MYLIB`. Your simulator documentation should provide guidelines on compiling design units into a library.

Step 7. Write a VHDL model of a bit-serial adder using signals of type **bit** and adopting the structure shown in Figure 6-9. The model must declare the library `MYLIB` provide access to your package via the **use** clause.

Step 8. Test the model of the bit-serial adder to ensure that it is functioning correctly.

Step 9. Modify the model to use signals of type `std_logic`. Nothing else should have to change including the structure of the procedure call. By virtue of the argument type in the procedure call, the correct procedure in the package that you have written will be used.

Step 10. Modify the **use** clause to limit the visibility to one procedure in the package. Repeat your simulation experiments. You might have multiple instances of the **use** clause to provide visibility to each of the procedures you wish to utilize.

Step 11. Repeat the experiment to use other models of the D flip-flop that include signals such as reset and clear.

End Simulation Exercise 6.1

6.6 Chapter Summary

Designs can become large and complex. We need constructs that can help designers manage this complexity and enhance sharing of common design units. This chapter has addressed the essential issues governing the construction and use of subprograms: functions and procedures.   Commonly used subprograms can be organized into packages and placed in design libraries for subsequent reuse and sharing across distinct VHDL models. The concepts introduced in this chapter include the following:

- Functions
    - type conversion functions
    - resolution functions
- Procedures
    - concurrent procedure calls
    - sequential procedure calls
- Subprogram overloading
    - subprogram name
    - operator overloading
- Visibility rules
- Packages
    - package declaration
    - package body
- Libraries
    - relationships between design units and libraries

We are now armed with constructs for hiding complexity and sharing and reusing VHDL code modules.

Exercises

1. Create a package with functions–procedures for performing various shift operations and increment and decrement operations on **bit_vector** elements and `std_logic_vector` elements. Place the package declaration and package body in distinct files and analyze them separately. Remember to declare and use the library `IEEE` and the package `std_logic_1164.vhd`.

2. Consider a VHDL type that can take on the values (0, 1, X, U). The values X and U correspond to the values unknown and uninitialized respectively. Define a resolved type that takes on these values and write and test a resolution function for this resolved type.

3. Write and test a set of procedures for performing arithmetic left and right shifts on vectors of type `std_logic_vector`.

4. Write and test a resolution function that operates on elements of type `std_logic_vector` and returns the largest value.

5. Write and test functions that can perform type conversion between multibit quantities of type `std_logic_vector` and integers.

6. Using a concurrent procedure looks very much like using a component in a hierarchically structured design. What is the difference between using a concurrent procedure and constructing a structural design?

7. Create a design library `My_Lib` and place a package in this library. You might create a package of your own or simply "borrow" any one of a number of existing packages that come with VHDL environments. Analyze this package into this library. The creation of this library with the logical name `My_Lib` will involve simulator specific operations. Ensure that you have correctly implemented this library by using elements of this package in a VHDL model analyzed into your working library, `WORK`.

CHAPTER 7   Basic Input/Output

Thus far, we have written VHDL programs that manipulate three classes of objects: variables, signals, and constants. We adhere to certain rules when using these objects. For example, if we were to use a variable in our program, we would give it a name such as `Index`, and we first declare the type of `Index` using a declaration statement. The range of values that the program can legally assign to `Index`, and the operations that can be performed on `Index` are determined by its type. For example, if `Index` is an integer, we may perform integer arithmetic using `Index`, and the range of values that it may take depends on the number of bits used to represent an integer: 16, 32, and so forth. In an analogous fashion, the use of input/output functions necessitates the introduction of the *file* type that permits us to declare file objects.

Files are special and serve as the interface between the VHDL programs and the host environment. They are manipulated in a manner very different from variables or signals, hence the need for a distinct object type. As you might expect, there are special operations that are performed only on files: reading and writing files. This chapter discusses how file objects are created, read, written, and used within VHDL simulations. A very useful example of the application of file I/O is the construction of testbenches—VHDL programs for testing VHDL models. The notion of a testbench follows directly from the testing of chips and boards and provides a structured approach for validating designs captured in VHDL models.

7.1 Basic Input/Output Operations

As with variable, signal, and constant objects, before we can use a file object we must give it a name and declare its type. Files can be distinguished by the type of information that can be stored within a file. For example, a file of characters can be thought of as a file of type `text`. Similarly, a file containing integers can be thought of as a file of type `INTF`. We can declare a file type with the following type declaration.

> **type** `TEXT` **is file of** `string`;

The preceding declaration defines a *file type* which can store ASCII data. Files of this type contain human readable text. Now that we can produce files of a certain type we need procedures and functions to read and write files of this type. These are usually provided by bundling them in the form of packages. For example, the package `TEXTIO` provides procedures and functions to read/write files of type `TEXT`.

Recall that packages can be thought of as a library or repository of predefined types, variables, signals, constants, functions, and procedures. The `TEXTIO` package is distributed with VHDL simulators. This package is placed in the library `STD` which is *implicitly visible*. This means that your model does not need to include a **library** clause declaring `STD` as a library. However, if you wish to use functions in the `TEXTIO` package a **use** clause must be included as we have seen for other packages. The package `TEXTIO` defines a standard file type called `TEXT` and provides the procedures for reading and writing the predefined types of the language, such as **bit**, **integer**, and **character**. CAD tool vendors will also often provide a set of library functions and procedures for input and output operations. Refer to the toolset documentation of the simulator that you are using for a description of other I/O functions that are supported.

From within a VHDL program we need some way to reference a file. This is performed via a file declaration which for VHDL'87 will appear as follows:

> **File** `infile` : `text` **is in** "`inputdata.txt`";

> **File** `outfile` : `text` **is out** "`outputdata.txt`";

Note the case of the file type `text`: VHDL is case insensitive. The object `infile` can be regarded as a pointer to a file to be opened for input. The name of the file is `inputdata.txt`. Since the file is open for input, the access mode of the file is said to be of type **in**. The default location of this file is the current working directory. Similarly, the object `outfile` is declared as a pointer to a file named `outputdata.txt` The access mode for this latter file is declared to be of type **out** and can only be written. When VHDL

VHDL'93 was revised in 1993 changes were introduced in the input and output operations. In contrast to VHDL'87, file declarations in VHDL'93 appear as shown below.

> **File** `infile` : `text` **open** `read_mode` **is** "`inputdata.txt`";

> **File** `outfile` : `text` **open** `write_mode` **is** "`outputdata.txt`";

The file objects `infile` and `outfile` can be used by VHDL procedures to read and write the files `inputdata.txt` and `outputdata.txt` respectively. What are these I/O procedures and how are they used? In conventional programming languages I/O operations involve calls to initialization procedures to open files *prior* to reading or writing data. Other procedures must also be called prior to closing files. For example, the C language provides the `fopen()` function call that returns a pointer to a file. The `fclose()` function call is provided to close the file and make sure the last updates have been written out to the file on disk (since they may still be cached in memory). In between calls to these two procedures, a file can be read or written using functions such as `fscanf()` and `fprintf()`. The arguments of these functions include a pointer to the file and the variables to be written or read. Most other programming languages provide equivalent functions to be used in a similar manner. In principle, the VHDL language operates in much the same way with a few important differences.

Let us assume that we have the predefined file type `TEXT`. Thus, we can read and write strings of characters. The standard VHDL procedures made available by the definition of the language are as follows.

procedure READ (**file** f : TEXT; value : **out** type);

procedure WRITE (**file** f : TEXT; value : **in** type);

function ENDFILE (**file** f : TEXT) **return boolean**;

The above I/O procedures are implicitly declared following a file type declaration. Whereas the procedures READ and WRITE are used for input/output operations, the END-FILE function is used to test for the end of file when reading from files. Most VHDL simulators will provide a set of procedures for reading and writing various data types: **character**, **integer**, **real**, and so on. Refer to your simulator documentation for more details on the available I/O procedures for reading and writing different data types. The standard package TEXTIO provides procedures for reading and writing files of type TEXT. More information on the package TEXTIO is provided in Section 7.2.   **VHDL'93**

In VHDL'87 files cannot be opened or closed. In VHDL'93 the following two functions are also provided and implicitly declared following a file type declaration.

procedure FILE_OPEN (**file** f : TEXT; External_Name : **in** STRING;

Open_Kind : **in** FILE_OPEN_KIND := READ_MODE);

procedure FILE_CLOSE (**file** f : TEXT);

The arguments specify a pointer to the file, a filename, and the mode—input or output. Normally, you would expect that the application programmer must make a call to FILE_OPEN prior to any use of the file and a call to FILE_CLOSE after the last operation on the file. If we use the preceding file declaration, then the FILE_OPEN is not necessary since the declaration specifies the mode of the file. An implicit call to FILE_CLOSE exists and is called when execution terminates. These declarations are useful if we were performing more advanced I/O operations and were not using the text I/O facility distributed with VHDL simulators. In this text we introduce the basic text I/O operations while these more advanced features can be found in many excellent VHDL texts [2,10].

```vhdl
library IEEE;
use IEEE.std_logic_1164.all;
use STD.textio.all;

package classio is
procedure read_v1d (variable f: in text; v : out std_logic_vector);
procedure write_v1d (variable f:out text; v : in std_logic_vector);
end;

package body classio is
procedure read_v1d (variable f:in text; v : out std_logic_vector) is
variable buf: line;
variable c : character;

begin
readline (f, buf);
for i in v'range loop
read(buf, c);
case c is
  when 'X' => v (i)  := 'X';
  when 'U' => v (i)  := 'U';
  when 'Z' => v (i)  := 'Z';
  when '0' => v (i)  := '0';
  when '1' => v (i)  := '1';
  when '-' => v (i)  := '-';
  when 'W' => v (i)  := 'W';
  when 'L' => v (i)  := 'L';
  when 'H' => v (i)  := 'H';
  when others => v (i)  := '0';
end case;
end loop;
end;

procedure write_v1d (variable f: out text; v : in std_logic_vector) is
variable buf: line;
variable c : character;

begin
for i in v'range loop
case v(i) is
```

FIGURE 7-1 An example of using character I/O to read and write single-bit vectors

```
      when 'X' => write(buf, 'X');
      when 'U' => write(buf, 'U');
      when 'Z' => write(buf, 'Z');
      when '0' => write(buf, character'('0'));
      when '1' => write(buf, character'('1'));
      when '-' => write(buf, '-');
      when 'W' => write(buf, 'W');
      when 'L' => write(buf, 'L');
      when 'H' => write(buf, 'H');
      when others => write(buf, character'('0'));
    end case;
    end loop;
    writeline(f, buf);
    end;
    end classio;
```

FIGURE 7-1 (cont.)

Example: Basic Input/Output Operations

Let us suppose that we have only the predefined file type of TEXT. Such files contain only character strings. However, suppose that we wish to read and write binary signal values of the type std_logic_vector. This example illustrates how we can develop input/output procedures to read and write binary signal values based on available elementary character I/O operations that are provided by VHDL. If we can provide these new procedures in a package, we can then use these procedures in our VHDL models and hide the fact that we really have only character I/O. This is exactly what most VHDL environments do by providing a set of functions in the package TEXTIO to read various data types. Note that we do need the **use** clause so that the contents of the TEXTIO package is visible to the VHDL programs.

Figure 7-1 shows a package with two procedures. The first procedure is for reading a bit vector of type std_logic_vector and the second is for writing a bit vector of type std_logic_vector. Each procedure requires a file pointer as an argument. Each procedure also accepts as a parameter a variable v of type std_logic_vector. We can use the same procedures for reading and writing bit vectors of differing precision. Note how each procedure works. In the read procedure a complete line is read into the buffer buf using the **readline**() VHDL procedure call. This procedure is defined in the package TEXTIO that is contained in the library STD. This library and package exist in all VHDL environments. Once the complete vector is read into buf, this buffer is scanned to test for the value of each digit in the vector and to set the value of each output digit accordingly using the case statement. Finally, note that the loop scanning the input vector executes a number of times that is determined by the **'range** attribute of the vector being read. An analogous procedure is used for writing bit vectors of type std_logic_vector out to a file. Note that the variable buf can be accessed only via the **read** () and **write** () procedures. This is because buf is of type line which is a special type referred to as an **access** type. An access type is very similar to a pointer in C or Pascal. While providing powerful programming flexibility we are more concerned here with the hardware modeling aspects of VHDL. Therefore other than its use in file I/O, we will not deal with access types any further. Since we are dealing with text-based I/O, when writing the bit values 0 and 1, they must be first converted to characters as shown in write_v1d() in the figure. This is necessary so that the implementation may distinguish between the value of 0 and 1 and its corresponding ASCII character representation. By using the basic read and write procedures provided by VHDL, we are now able to read and write single-bit vectors of the type std_logic_vector. The definitions and procedures defined in the TEXTIO package are shown in Appendix D. You will see that the **read** () and **write** () procedures are overloaded

Example End: Basic Input/Output Operations

We can similarly construct procedures for reading and writing other data types from files. Generally, rather than the user developing these procedures, the CAD tool vendor will provide a comprehensive set of input/output functions for reading and writing various data types. For example, the Viewlogic Systems Workview Office toolset makes available a package called pack1076, which contains input/output procedures for reading and writing a wide range of data types.

Example: Using the Classio Package

An example of how the procedures and the package shown in Figure 7-1 might be used is shown in the example code in Figure 7-2. This example simply reads 16-bit vectors from infile.txt and writes them out to the file outfile.txt. Note that by changing the resolution of the variable check to 32 bits, the same procedures can be used for reading and writing 32-bit vectors. This is because the loop in the read/write procedures that scans the vectors is parameterized by the range of the vector as opposed to a fixed value. The simple test program is written using a wait statement to ensure that the simulation progresses. Remember that, in general, processes are executed once, when the simulation is initialized. Thereafter, processes are executed only if they are invoked by events. Events may occur in a sensitivity list or via the use of wait statements. In general, you will proba-

```
library IEEE;
use IEEE.std_logic_1164.all;
use STD.textio.all;
use work.classio.all;

entity checking is
end checking;

architecture behavioral of checking is
-- VHDL 1987 Example
begin
process
-- declare pointers to the input and output files
file infile : TEXT is in "infile.txt";
file outfile : TEXT is out "outfile.txt";

variable check : std_logic_vector (15 downto 0) :=
to_stdlogicvector(x"0008");

begin
-- copy the input file contents to the output file
wait for 10 ns;
while not (endfile (infile)) loop
read_v1d (infile, check);
write_v1d (outfile, check);
end loop;
end process;
end behavioral;
```

FIGURE 7-2 An example of the use of the package *classio.vhd*

bly be performing I/O operations within some process that will be invoked due to some simulation events. The use of the wait statement in the above example is somewhat artificial.

Example End: Using the Classio Package

Simulation Exercise 7.1: Basic Input/Output

This exercise introduces the student to the basic steps involved in reading and writing text files, covering the following concepts: (i) declaring, and opening files, (ii) initialization of data structures from files, and (iii) recording simulation results in file.

Step 1.   Create a text file, *memory.vhd*, with the model of a memory module from Figure 4-2. The succeeding steps will modify this model to initialize the memory to encoded values read from a file.

Step 2.   Create the file *classio.vhd* shown in Figure 7-1. This is a package. Compile this package into your working directory. The simulator you are using should have the default library WORK set to this directory.

Step 3.   Edit *memory.vhd* to enable references to the package *classio.vhd* by placing the following clause at the beginning of the file.

   use WORK.classio.all;

Step 4.   Create a text file *infile.txt*. This file should include the contents of memory. Each line should be a 32-bit vector of type std_logic_vector, and there should one word per line. For the model we are using, the file should have eight words.

Step 5.   Add a single-bit signal, reset, as an input signal to the entity description. This signal is of mode **in** and of type std_logic. A pulse on this signal will indicate that the memory contents are to be initialized from a file.

Step 6.   Add a process to the model to read eight 32-bit values from a file and initialize the contents of memory to these values.

   Step 6 (a) Label this process ioproc.

   Step 6 (b) Have the first line in the process be a wait statement, causing the process to wait for a rising edge on the signal reset. Examples of the detection of rising edges on signals can be found in the D flip-flop examples in Chapter 4.

   Step 6 (c) Use the functions in the package classio in writing the body of the process. The process body reads and initializes the contents of memory from the input file. The model in Figure 7-2 can be used as an example of how a **while** loop can be used for this purpose.

Step 7.   The modified memory model is now complete. Compile the model into the working directory.

Step 8.   Test the model as follows.

   Step 8 (a) Create a sequence of test inputs to the model. For the reset signal we must supply a single pulse. For the remaining inputs, supply a sequence of addresses starting from 0 and through memory address 7. The memory read control signal must be asserted.

Step 8 (b) Select the signals to be traced. Trace all of the signals in the entity description.

Step 8 (c) Apply the stimulus as generated in Step 8 (a). Observe the trace.

Step 8 (d) After eight memory read operations the values read from memory should be identical to the values initialized from the file.

Step 9. Modify the model to use an output file *outfile.txt*. Now generate a sequence of read and write operations to memory using the external stimulus capabilities of the simulator you are using.

Step 10. Now modify the model to log all memory operations to the output file. For each read and write operation, the model records the value and address in *outfile.txt*.

Step 11. Verify that the contents of *outfile.txt* agree with the test sequence that you generated in Step 9.

Step 12. Modify the memory model so that data types used are **bit** and **bit_vector** rather than `std_logic` and `std_logic_vector`. For example, memory would be an array of words of type **bit_vector**. Now we can use the I/O functions in the package `TEXTIO`. These functions are shown in Appendix D. Modify the model to use the I/O functions from this package. Note that these function names are overloaded.

Step 13. Rerun the model using the `TEXTIO` package rather than the `classio` package. I/O procedures are available for all of the predefined types of the language, for example, strings. Modify the **use** clause accordingly. Remember, `std_logic` and `std_logic_vector` are not predefined types of the language. They are defined by the package `std_logic_1164`!

End Simulation Exercise 7.1

7.2 The Package TEXTIO

The package `TEXTIO` is a standard package supported by all VHDL simulators. This package provides a standard set of file types, data types, and input/output functions. In general, when you use a package in a VHDL model, you must declare the package to be used. The declaration will state the library (system directory) that serves as the location of the package. The `TEXTIO` package is in the library `STD`, which is implicitly declared. This means that unlike user libraries, you do not have to declare the use of `STD` as we do with the library `IEEE`. However, usage of the package contents must be declared via the **use** clause as shown in the preceding examples.

The `TEXTIO` package provides the definition of several data types and procedures. These are shown in the package definition provided in Appendix D. For example, the `text` object type and file handles for `std_input` and `std_output` are provided.

From Appendix D we see that **read** () and **write** () procedures are defined for several data types. Normally, we expect that each data type will have a distinct procedure for reading and writing the type to a file. For example, we could have `read_string` () and `read_bit_vector` (). However, given the large number of data types, we see that this quickly becomes a rather tedious exercise in naming procedures. Instead, it is convenient for all of the procedures that perform the same function, such as **read**(), to have the same name, and permit the actual argument to identify the correct implementation: **read** () and **write** () are overloaded procedure names. We do not have to remember the exact name of the procedure to read and write elements of type **bit_vector** or **string**. We simply use the procedures **read** () and **write** (), and, depending on whether the argument is a **bit_vector** or **string** the appropriate implementation is invoked. We see that **read** () and **write** () procedures are available for the predefined types **bit**, **bit_vector**, **character**, and **string**. As illustrated in Figure 7-1, we can construct input/output procedures for other data types using these basic procedures to operate on more complex data types. Check the your CAD vendor documentation for supported run-time procedures and functions.

7.3 Testbenches in VHDL

Much of the motivation for simulation is to be able to test designs prior to construction and use of the circuit. How would we test an electronic component? Intuitively, we would like to provide a set of inputs and check the observed output values against the corresponding correct set of output values. The component can be placed in a specialized piece of equipment, referred to as a test frame, that allows us to apply an electrical stimulus to the inputs and examine the values of the outputs using a logic analyzer. By cycling through possible input sequences, we could analyze the corresponding output sequences to determine if the component was functioning correctly.

When developing VHDL models, we find ourselves in a similar situation. We construct a VHDL simulation model of some digital system component such as an encoder for audio signals. How do we test this model to ensure that (i) the model is operating as designed, and (ii) the design itself is correct? Verifying a design via simulation is certainly more cost-effective than testing a fabricated part and then determining that it has a design error that must be fixed. If our VHDL model is sufficiently detailed, then thorough testing in simulation significantly reduces the chances of errors, minimizes costly design rework, and reduces the time to get the product to the marketplace.

Most simulators provide commands to apply stimulus to the input ports of a design entity. By tracing and viewing the resulting values of the signals on the output ports, we can determine whether the model is operating correctly. However, by recognizing that VHDL provides powerful programming language abstractions for describing the operation of digital systems a more structured approach toward testing VHDL models can be realized. The test frame described above is itself a digital system and we should be able to describe the operation of the tester in VHDL! The logical behavior of the tester is simple to understand. The model generates sequences of inputs and reads the outputs of the mod-

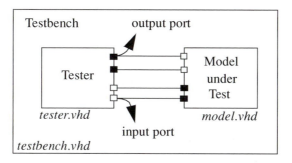

FIGURE 7-3 The structure of a testbench

ule being tested. This behavior is captured in the notion of a *testbench* and is illustrated in Figure 7-3.

In this figure, one VHDL module (*tester.vhd*) generates the stimulus to be applied. A second VHDL module (*model.vhd*) is the model being tested. Finally, a third module (*testbench.vhd*) is a structural VHDL model that describes the interconnections between the tester and the model under test. This model describes how the ports of the tester are connected to ports of the model. The simulation progresses with the tester applying a stimulus to the model and reading and recording the responses. The sequences of input values applied to the input ports of the model under test can be read from a file, or they can be generated using VHDL constructs. For example, Section 4.6 describes some approaches to generating waveforms. The results returned from the VHDL model can be checked by the tester against a known correct set of output values and errors can be flagged. When testing large systems with a potential of millions of test vectors, automation of this process is a very productive enterprise. The construction of testbenches is best illustrated with an example.

Example: Writing a Testbench

Let us assume we wish to write a testbench to test the VHDL model of the positive edge-triggered D flip-flop developed in Chapter 4 and shown in Figure 7-4. In order to test this model, we must generate a clock signal and a sequence of values on the D input signals. We must also test the Set (S) and Reset (R) operations. A sample test pattern for the clock and input (D) is shown in Figure 7-5. The D input is sampled at the rising edge of the clock producing the output waveforms shown in the figure. The Set (S) input and Reset (R) inputs can be tested by applying a test vector where either S or R or both are asserted and then reading the value of the flip-flop outputs after a delay equal to the propagation delays of the signals through the flip-flop.

```
library IEEE;
use IEEE.std_logic_1164.all;

entity asynch_dff is
port (R, S, D, Clk : in std_logic;
      Q, Qbar : out std_logic);
end asynch_dff;

architecture behavioral of asynch_dff is
 begin
output: process (R, S, Clk)
 begin
 if R = '0') then
    Q <= '0' after 5 ns;
    Qbar <= '1' after 5 ns;
 elsif S = '0' then
    Q <= '1' after 5 ns;
    Qbar <= '0' after 5 ns;
  elsif (Clk'event and Clk = '1') then
    Q <= D after 5 ns;
    Qbar <= ( not D) after 5 ns;
 end if;
 end process output;
 end behavioral;
```

FIGURE 7-4 Behavioral model of a positive edge-triggered D flip-flop

An example of a tester module that generates the clock signal and applies a set of test vectors to the D flip-flop model is shown in Figure 7-6. Note that the example I/O package shown in Figure 7-1 is compiled into the library WORK. Consider the structure of the tester module. The process named clk_process generates a clock signal with a period of 20 ns. This process executes concurrently with the io_process. This latter process reads test vectors from file *infile.txt*. Each test vector is five bits long. The first three bits correspond to input values for R, S, and D. The last two bits are the corresponding correct values of Q and Qbar. Collectively these five values constitute one test vector. The input test vector values are applied to the flip-flop model at 20 ns intervals—the rising edge of the clock. From the correct timing behavior shown in Figure 7-5 we can generate a set of test

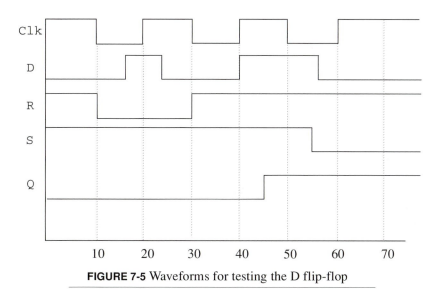

FIGURE 7-5 Waveforms for testing the D flip-flop

vectors by looking at the values of the inputs on each rising edge (i.e., at time 20 ns, 40 ns and so on). This can give rise to the following test vectors in file *infile.txt*.

    11001   -- initial vector
    01101
    11110
    10010
    10011   -- illegal case

Each vector is applied at 20 ns intervals to coincide with the rising edge of the clock. The last test vector corresponds to an illegal case, since both Q and Qbar cannot be asserted at the same time. Therefore this test vector will fail and an error message will be logged to the output file *outfile.txt*. This example also shows that failure of a test vector may not be due to an incorrect model but an incorrect test vector. The generation of test vectors in general is a computationally intensive and non trivial task. Once the stimulus is applied, the process waits for 20 ns. By this time the flip-flop outputs become stable (see Figure 7-4). At this time, the outputs are read and compared to known correct values that were read from the input file. If there is any discrepancy in these values, the corresponding test vector is flagged by writing out an error message to an output file along with the test vector. After all the test vectors have been read and applied, the output file contains the list of test vectors for which the model failed.

The top-level module for our testbench is a structural model for specifying the connections between the tester and the model under test and is shown in Figure 7-7. Note the use of configuration specifications to explicitly state which VHDL architectures are to be

```vhdl
library IEEE;
use IEEE.std_logic_1164.all;
use STD.textio.all;
use WORK.classio.all; -- declare the I/O package

entity srtester is                -- this is the module generating the tests
port (R, S, D, Clk : out std_logic;
    Q, Qbar : in std_logic);
end srtester;

architecture behavioral of srtester is
begin
clk_process: process  -- generates the clock waveform with
begin                          -- period of 20 ns
Clk<= '1', '0' after 10 ns, '1' after 20 ns, '0' after 30 ns;
wait for 40 ns;
end process clk_process;

io_process: process                   -- this process performs the test
file infile : TEXT is in "infile.txt";        -- functions
file outfile : TEXT is out "outfile.txt";
variable buf : line;
variable msg : string(1 to 19) := "This vector failed!";
variable check : std_logic_vector (4 downto 0);

begin
while not (endfile (infile)) loop    -- loop through all test vectors in
read_v1d (infile, check);        -- the file
R <= check(4);
S <= check(3);
D <= check(2);
wait for 20 ns;                -- wait for outputs to be available after applying
                               -- the stimulus
if (Q /= check (1) or (Qbar /= check(0))) then -- error check
write (buf, msg);
writeline (outfile, buf);
write_v1d (outfile, check);
end if;
end loop;
wait;     -- this wait statement is important to allow the simulation to halt!
end process io_process;
end behavioral;
```

FIGURE 7-6 Behavioral description of the tester module

```
library IEEE;
use  IEEE.std_logic_1164.all;
use WORK.classio.all; -- declare the I/O package

entity srbench is   -- the entity interface is empty
end srbench;

architecture behavioral of srbench is
component asynch_dff
port (R, S, D, Clk : in std_logic;
     Q, Qbar : out std_logic);
end component;

component srtester
port (R, S, D, Clk : out std_logic;
     Q, Qbar : in std_logic);
end component;
--
-- configuration specification
--
for T1:srtester use entity WORK.srtester (behavioral);
for M1: asynch_dff use entity WORK.asynch_dff (behavioral);

signal s_r, s_s, s_d, s_q, s_qb, s_clk : std_logic;

begin
T1: srtester port map (R=>s_r, S=>s_s, D=>s_d, Q=>s_q, Qbar=>s_qb,
                     Clk => s_clk);
M1: asynch_dff port map (R=>s_r, S=>s_s, D=>s_d, Q=>s_q,
                       Qbar=>s_qb, Clk => s_clk);
end behavioral;
```

FIGURE 7-7 Structural description of the testbench module

used for the design entities srtester and asynch_dff. In general, there could have been more than one architecture that we could have used. The configuration specification states that the architecture labeled behavioral is to be used for each entity. The configuration also states that these architectures can be found in the working directory denoted by the library WORK. If no configuration specification had been provided, default rules would apply and the last compiled architecture for entities srtester and asynch_dff would have been used.

When testing combinational circuits there may be no clock or periodic signal. In this case the tester module can be written without having to be concerned about synchronizing

assert Q = check(1) **and** Qbar = check(0)
report "Test Vector Failed"
severity error;

FIGURE 7-8 An example of the use of the **assert** statement in a testbench

with a periodic signal. The module can apply the input vectors, wait for a period equal to the propagation delay of the longest path through the circuit, and then read the output signal values of the module under test.

Example End: Writing a Testbench

7.4 ASSERT Statement

In the prior example, the testbench model recorded errors or failures to pass a test vector by writing the test vector and a brief message to the file outfile.txt. Alternatively, we could use the **assert** statement. The **assert** statement is a general mechanism for detecting and reporting incorrect conditions during a simulation. We can use this statement for detecting failed test vectors, as shown in Figure 7-8. This statement would replace the **if** statement in the srtester module in Figure 7-6 that checks the output signal values and causes error messages to be written to a file. If we use the **assert** statement, rather than having an error message and the offending test vector being written to a file, the message regarding a violation and the string provided with the **report** clause would be sent to the simulation output. This output is generally the simulator console window unless you have redirected this to a file. The designer can report messages at one of several predefined severity levels: NOTE, WARNING, ERROR, and FAILURE. This provides a clean mechanism for the designer to classify the levels of information conveyed during simulation. For example, the NOTE category can be used to provide information about the progress of the simulation while a severity level of ERROR may cause the simulation to be aborted.

7.5 A Testbench Template

We are now ready to outline a few basic steps toward constructing a testbench for testing a VHDL model. It should be clear from the above examples that the testbench is a structural model with two components: a tester and the model under test. This is not the only way to structure a testbench, but is intuitive and simple to construct. Toward the end of the chapter we will describe an alternative structure for a testbench module.

The model under test may be a behavioral or structural VHDL model of a digital system. The tester is usually a behavioral model written using the constructs described in Chapter 4. Typical segments of VHDL code that we may find in tester modules include (i) processes to generate waveforms, (ii) VHDL statements to read test vectors from input files and apply them to the model under test, and (iii) VHDL statements to record the outputs that are produced by the model under test in response to the test vectors. A template for such a top-level structural model is shown in Figure 7-9. A step-by-step description for the construction of such a testbench model is the same as that for creating structural models as described in Chapter 5.

Recall from Chapter 4 that we can mix concurrent and sequential statements (via processes) in the architecture description of a circuit. Rather than have distinct models for the testbench module, tester module, and the model under test, the tester VHDL code may be directly included the testbench architecture. The component instantiation statement T1 in Figure 7-7 can be replaced by the tester code. If we took this approach we would have only two code modules: the testbench and the model under test.

Simulation Exercise 7.2: Constructing Testbenches

This exercise familiarizes the student with the basic steps involved in constructing and executing a testbench for testing and validating a VHDL model. Basic concepts covered include (i) generating a stimulus for a model under test, (ii) reading test vectors from a file and applying them to the VHDL model under test, (iii) recording failed test vectors and generating error messages for examination

Step 1. Using a text editor, create the files *dff.vhd*, *srtester.vhd*, and *testbench.vhd* shown in Figure 7-4, Figure 7-6, and Figure 7-7 respectively. These files should be in your working directory. Compile each file.

Step 2. Create the file *classio.vhd* shown in Figure 7-1. This is a package. Compile this package into your working directory. The simulator you are using should have the default library WORK set to this directory.

Step 3. Create a text file named *infile.txt*. This file contents should appear as follows:
    11001
    01101
    11110
    10010

This file represents a set of test vectors to be applied to the model *dff.vhd*. The first three bits (left to right) represent test inputs for the R, S, and D inputs respectively. The last two inputs represent the corresponding correct values for Q and Qbar, respectively.

Step 4. Create the empty text file *outfile.txt*.

Step 5. Open a trace window and select the signals you wish to trace.

library Lib1;   -- declare any libraries that will be needed
library Lib2;
use Lib1.package_name.all; -- declare the packages that will be used
use Lib2.package_name.all; -- in these libraries

entity test_bench_name **is**

port(*input signals* : **in** *type*;
     output signals : **out** *type*);

end test_bench_name;

architecture arch_name **of** test_bench_name **is**

-- declare tester and model components
component tester_name
port (*input signals* : **in** *type*;
     output signals : **out** *type*);
end component;

 component model_name
port(*input signals* : **in** *type*;
     output signals : **out** *type*);
end component;

-- declare all signals used to connect the tester and model

signal *internal signals* : *type* := initialization;

begin

-- label each component and connect it's ports to signals or other ports

T1: tester-name **port map** (port=> signal1,.....);

M1: model-name **port map** (port => signal2,.....);

end arch_name;

FIGURE 7-9 A testbench model template

Step 6.   Run the simulation long enough to apply the test vectors.

Step 7.   Examine the output file *outfile.txt* for any error messages. How many error message should appear, if any, for the above sequence of test vectors?

Step 8.   Modify some of the test vectors in *infile.txt*. Rerun the simulation and study the input and output files to determine if the model is functioning correctly.

Step 9.   Modify the tester module as follows: consider the **if** statement in *srtester.vhd* that compares the output of the D flip-flop with the test vector and writes *outfile.txt*. Replace this statement with an ASSERT statement as follows.

> **assert** Q = check(1) **and** Qbar = check(0)
> **report** "Test Vector Failed"
> **severity error**;

Step 10.   Rerun the simulation. Any previous error messages should appear as an assertion violation accompanied by the error message being printed on the simulator command line rather than being directed to the file *outfile.txt*.

End Simulation Exercise 7.2

The models we are dealing with here are relatively small. We can see that as models become more complex the number and size of the test vectors could become very large. For example, consider a circuit that processes 32-bit data according to a 16-bit operation code. At the very least, this circuit will have 48 inputs. A naive testing approach would attempt to test all possible input combinations. The total number of possible combinations of input values (assuming only 0 and 1 values) is 2^{48}! Even if we could perform each test in a nanosecond it would still take on the order of days to finish the test suite. Furthermore, the number of inputs in modern chips and systems is considerably higher. In reality, the number of input combinations that must be tested can be pared considerably and many computational as well as design techniques have been implemented to further reduce the cost of testing. Even so, it is apparent that the generation of test vectors can be a very complex process in its own right. To facilitate sharing of test vectors among groups (e.g., the people generating them and the designers using them), standards are often defined for the specification of these vectors.

Finally, remember that often input or output functions are expected to be executed only once, such as in loading memory in a processor datapath model. If the I/O code is placed in part of a process that is executed repetitively, array out of bounds errors can occur.

7.6 Chapter Summary

This chapter has presented the basic concepts for reading and writing files. One common application of file I/O was developed: writing testbenches for testing VHDL models. The exercises stressed basic text I/O from files, and construction of functions for reading and writing other data types. This concepts covered in this chapter include the following:

- Files, file types, and file declarations.
- Basic operations for reading and writing character strings from text files.
- Creation and use of procedures for reading and writing other data types.
- Construction and operation of testbenches.

Armed with an understanding of the basic issues of file I/O we can proceed to ask the right questions to determine the I/O functions that are supported within any commercial simulator. This includes declaring and using any general I/O packages that may be publicly available, such as the TEXTIO package, or creating a new package as required to read and write application specific data types.

Exercises

1. Write and test a model for a 16-bit shift register that is initialized to a value read from a file.

2. In models of complex components such as memories in modern processors, accesses to memory must follow certain restrictions. For example, you generally cannot write to the area of memory that stores program instructions. Modify the memory model used in Simulation Exercise 7.1 to include an **assert** statement with a severity level of NOTE. Use this statement to produce a message whenever a particular memory location is written. Compile and test the model.

3. What are some of the advantages to using the idea of testbenches rather than testing your model by providing stimuli to the entity input using the facilities in a simulator?

4. Consider a testbench for a combinational circuit such as the single-bit ALU described in Simulation Exercise 3.2. Write a VHDL model for this circuit. Develop a testbench to test the model. Verify the functionality of the ALU using this testbench.

5. Modify the D flip-flop model shown in Figure 7-4 to include the use of an **assert** statement to check for conditions when both the Set and Reset inputs are asserted.

6. Write a testbench and test a model for an 8-bit counter. Use **bit** and **bit_vector** types. Therefore you can use the procedures in the package TEXTIO to read test vectors from a file.

CHAPTER 8

Programming Mechanics

The creation and simulation of VHDL models is structured around a well-defined set of conventions. This chapter discusses the general conventions used to organize the simulation environment and VHDL programs. We cover the mechanics of organizing, building, and simulating VHDL models and the relationship to various programming constructs. Understanding these conventions is useful in debugging and reasoning about model behavior as well as quickly coming up to speed in being able to productively use VHDL-based simulators. This chapter provides an intuition about the practical aspects of VHDL simulator environments and hopefully eases the transition into proficient use and avoids some of the frustrations that often accompany navigation through CAD tool environments.

Just as we have compilation, linking, and loading of Pascal programs the major concepts in programming mechanics can be identified as *analysis, elaboration, initialization,* and *simulation.*

8.1 Analyzing VHDL Programs

Analyzing VHDL programs is synonymous with compiling VHDL programs. The two terms are often used interchangeably. Consider the process of compiling and executing a Pascal program. We start with a text file containing the program. This program may reference functions or procedures found in other program modules, such as function libraries created by the user, or system libraries that are a part of the programming environment.

For example, procedures to read and write files as well as many mathematical functions are provided by the system libraries while the user may have created a library of functions for manipulating data such as strings or image data. Common programming environments provide rules for referencing these libraries, and for compiling and linking independently compiled program modules to create a single executable program image. Similar conventions exist for compiling and linking distinct VHDL modules into a single image that can be used for simulation. The structure of the programming units and the concepts that govern their compilation and linking are similar to those governing conventional programming languages.

The   basic unit of VHDL programming is a *design unit*. A design unit is a component of a VHDL design and is one of

- Entity

- Architecture

- Configuration

- Package Declaration

- Package Body

Design units are contained in *design files,* which contain the VHDL source for the individual design units. A single design file may contain the source for one or more design units. Design units are *analyzed* to produce a form that can be used by a simulator. The *VHDL analyzer* performs the customary syntactic checks and compilation to a form executable by a VHDL simulator. This process is analogous to the process of compilation of conventional programs such as Pascal. The analyzed design units are placed in a *design library.* Design libraries are typically implemented as directories in most computing environments. A design library is identified by a logical name. This logical name is used to reference the library in a VHDL program. The relationships of all of the components are illustrated in Figure 8-1. The figure shows the libraries with logical names WORK, STD, and IEEE. The library WORK is the current working library. If a user package is referenced in a user design unit, the package must be analyzed and the compiled design unit placed in a designated library. The default library for the placement of the analyzed package is the library WORK.

An arbitrary number of libraries can exist and a given design can reference design units in one or more of these libraries as needed. Libraries may have been created by the designer or there may be libraries created by the VHDL tools vendor. There are two special libraries that exist in all VHDL environments. These are the libraries STD and WORK. Recall that a library is a directory. The STD library contains the analyzed (i.e., compiled) descriptions of two packages; STANDARD and TEXTIO. The STANDARD package contains the definitions of the predefined types and functions of the language. For example, the definitions of integer, real, and bit_vector types and functions for operations on these types are defined in the package STANDARD. The TEXTIO package contains the predefined types, functions, and procedures for reading and writing from files. The library WORK corresponds to your working directory. Analyzed design files are placed in the working directory. The logical names STD and WORK are defined by the system imple-

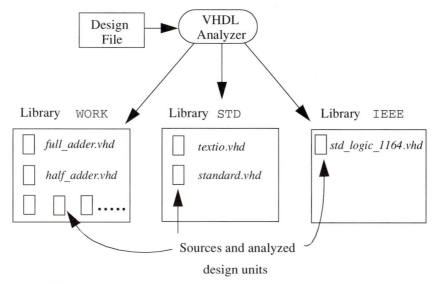

FIGURE 8-1 Placement and organization of design units and design libraries

mentation. This means that if your wish to use packages that are available in these libraries, you can do so via the **use** clause. You do not need to declare these libraries using the
library clause as we do with the library IEEE. Typically, each design tool vendor will
provide initialization files that assigns directory path names to STD and WORK when you
first invoke the VHDL environment. If there are other libraries present that you wish to use
in your VHDL programs, you must decide on a logical name to reference this library. This
logical name then must be setup in the host environment to point to the physical directory
that will contain the packages that constitute the library. Check your simulator documentation for details.

Example: Workview Office Setup

When using Viewlogic Workview Office 7.1, under Windows 95, the Speedwave simulator will require the creation of an initialization file that records the location of all logical
libraries. For example, the current working directory will contain a file created by the user
called *vsslib.ini*. This file will contain one line for each library used in the model. For
example, consider a model that uses the IEEE 1164 package stored in a library called
IEEE. Let the name of the library corresponding to the current working directory be *user*.
The following two lines will appear in this file:

```
D:\project\user.lib
D:\WVOFFICE\V\PGM\LIBS\IEEE.lib
```

In directory `\project`, `user.lib` will be found to be a directory that will contain the results of the analysis of various design units.

Example End: Workview Office Setup

8.2 Order of Analysis

Finally, with respect to analyzing VHDL programs, we must be concerned with the order in which design units are analyzed because of dependencies between them. Throughout the examples in this text, entities and architectures were maintained in the same file for convenience. This is not strictly necessary. Design units can be in independent files and analyzed separately. What then is the order in which a large set of design entities must be analyzed? What dependencies must we be aware of?

We can generate an intuition about compilation order by remembering that VHDL is a hardware description language. Consider the architecture of the structural model of a full adder reproduced from Figure 5-2 and shown here in Figure 8-2. In order to build the circuit shown we must have first "built," so to speak, the half-adder circuits and the two-input OR gate. Analogously, we must first have analyzed these design entities before we can analyze the architecture named `structural`. In general, for hierarchically structured models, we must remember to analyze hierarchical descriptions in a bottom-up fashion, starting from models at the lowest level in the hierarchy and proceeding to higher levels.

When we make changes to a design, we must consider dependencies between design units in determining which ones must be reanalyzed. Again let us pursue the hardware analogy. Suppose we describe a board level design by the interconnection of the component chips and the chip interfaces. If we now decide to replace one of the chips with a new chip with a different interface then we must take a closer look at the design and reanalyze it to make sure it is still functionally and electrically correct. On the other hand if we replace a chip with a newer, cheaper version, but one which maintains the same interface, we really do not have to perform any analysis and the design should remain correct. Similar relationships carry over to the relationships between VHDL design units. Consider the structural model in Figure 8-2. This model depends on the availability of an entity named `half_adder`. If that entity description is changed (as in changing the chip interface) then the `full_adder` model which *depends* on it must be reanalyzed. Thus, if any entity description is changed then *all* architectures that depend on that entity description must be reanalyzed. For similar reasons, if a package declaration is changed, then any design unit (entity, architecture, architectures) that depends on that package must be reanalyzed. This chain of dependencies may pass through multiple design units.

However, suppose all of the entity descriptions and architecture descriptions are in separate files. What if we now change the architecture and not the entity description? Continuing with the natural hardware analogy, consider the architecture in Figure 8-2. Suppose we wish to change this architecture to use some new, detailed two-input OR gate

```
architecture structural of full_adder is
component half_adder
port (a, b : in std_logic;
    sum, carry : out std_logic);
end component;

component or_2
 port (a, b : in std_logic;
     c : out std_logic);
end component;

signal s1, s2, s3 : std_logic;

begin
H1: half_adder port map (a => In1, b => In2, sum => s1, carry=> s3);
H2: half_adder port map (a => s1, b => c_in, sum => sum, carry => s2);
O1: or_2 port map (a => s2, b => s3, c => c_out);

end structural;
```

FIGURE 8-2 Structural model of a full adder

component model named fast_OR2. The half-adder model is not affected, nor is the entity description of the full adder which happens to be in a separate file. We simply need to (i) declare the component fast_OR2, (ii) change the component instantiation statement O1 to reflect the use of this new component, and (iii) reanalyze the architecture structural. Alternatively, suppose that the architecture of the half adder has changed but the entity description of the half adder remains the same. In this case, we simply reanalyze the architecture of the half adder. We do not need to reanalyze the full-adder model. From our hardware analogy, it is as if we had wired up the full-adder circuit and simply swapped out the half-adder chip for another one. As long as the entity description of the half adder has not changed, the changes in the architecture of the half adder are not visible to the full adder at this time. When the full-adder model is loaded into a simulator then the later architecture for the half adder will be used.

At this point in our VHDL modeling experience we may be inclined to organize our models with entity and architecture descriptions in the same physical file. Therefore, even if we only change the architecture description when we reanalyze the file the entity description is also reanalyzed and thus appears to have changed! All models that use this entity now have to be reanalyzed! This is because the environment determines if a design unit has changed by looking at the time stamp of the files created by the analyzer. If we reanalyze an entity description, even if we have not changed the interface the creation time of the analyzed files will have changed. The VHDL environment must operate on the assumption that the entity description has changed. We now cannot simulate any model that uses this entity without reanalyzing it.

In summary, analyze design units in a bottom-up fashion. When changes are made to entities or package declarations, all design units that depend on these units must be reanalyzed. If changes to the architecture or package bodies are made, then only these units need be reanalyzed. If only the architectures are to be analyzed, ensure that they are in separate files from the associated entity description.

8.3 Elaboration of VHDL Programs

We have seen that structural models are a means for managing large designs. Before a design can be simulated, it must first be flattened into a description of the system that can be simulated. We know that a concurrent signal assignment statement is a process. Now we see that flattening a design essentially produces a large number of processes that communicate via signals. These signals are referred to as *nets*. It is easy to think of a combinational circuit as a set of nets. Consider the gate level circuit shown in Figure 8-3. The signals s1, s2, s3, s4, s5, s6, and z, correspond to nets. A *netlist* is a data structure that describes all of the components connected to each net. Many industry standard formats exist for describing a netlist so that designs may be transferred easily between various design tools. This process of flattening a hierarchical description of a design is done during the phase of *elaboration* of the VHDL model. The elaboration produces a netlist of processes. Other functions are also carried out during elaboration and the overall process is comprised of the following steps.

1. *Elaboration of the Design Hierarchy:* This step includes flattening the hierarchy. In doing so, components must be associated with the architecture that is to be used to describe their behavior. This may be specified through a configuration or through the

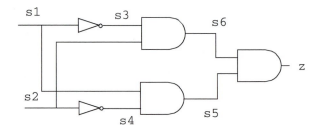

FIGURE 8-3 Netlists in a circuit

default choice of the architecture for an entity. Flattening the hierarchy produces a netlist of processes where each process describes the behavior of each component at the lowest level of the hierarchy.

2. *Elaboration of the Declarations*: Recall that generic parameters must be constants and their values must be known prior to simulation. These values are determined and checked in this step. The declarations are also checked for type consistency and initialization of signals and variables. The language has many rules that govern the initialization of signals and variables. These checks are completed in this step.

3. *Storage Allocation*: Storage is allocated for variables, signal drivers, constants and other program objects.

4. *Initialization*: All signals and variables are initialized in this step, either to user specified values such as those provided in declarations or to default values.

8.4 Initialization of VHDL Programs

Prior to simulation two important actions take place. First, all signals (nets) are initialized to their default or explicitly initialized values. Second, all processes are executed until they are suspended explicitly by the use of wait statements or implicitly by the use of a sensitivity list. This execution may also produce values for signals. Simulation time is set to 0 ns, and the model is ready to be begin simulation.

8.5 Simulation of VHDL Programs

Simulation proceeds as a discrete event simulation of the design. This is achieved by evaluating the values of all signals. If an event has occurred on any signal (i.e., net), all of the processes affected by that signal are executed, possibly generating future events on other nets. The events are conceptually managed as an event list organized according to the time stamp of the event as described in Chapter 2.

Let us consider what actually happens when a structural model such as the one shown in Figure 8-2 is simulated. We might first create a file with this model. Let us call this file *structural.vhd*. The models for the half adder and the two-input OR gate may be created in two other files, say *ha.vhd* and *OR2.vhd*. We now know that the order of compilation is important The latter two files are first analyzed. Independent of the file names, the compiled design units will be identified by their entity and architecture labels.

The process of simulation utilizes a number of concepts that are realized in various simulators in different ways. Some of the common steps that you can expect to encounter are the following:

8.5.1 Initialization

The simulator environment must maintain information about various design units involved in simulation such as the location of libraries. We know that libraries are logical names for directories. Most simulators need access to information about the libraries that you plan to use (such as IEEE), the location of your working design library (WORK), and the location of the library containing the standard packages expected with the VHDL distribution (STD). This information is usually created and maintained in initialization files. The files may be set by editing default files created at installation time, or they may be explicitly created when you use the simulator for the first time.

8.5.2 Loading the Model

For the example in Figure 8-2, the compilation will produce a design unit named full_adder. Simulators will provide facilities for loading a model. When the design unit full_adder is loaded into a simulator, the working directory (WORK), IEEE, STD, and any other libraries that you may have declared will be searched for any compiled design units with the labels half_adder and or_2. The order in which these libraries are searched is important, since you do not want to inadvertently use a design entity of the same name in another library. This order of search is simulator specific although it is not uncommon for the order to be based on that in which they are listed in the initialization files. Usually you would want the library WORK to be the first in the search order.

In this example, since no configuration is explicitly provided, the simulator will look for a design unit with the same name as the component and an architecture associated with *that* component. In our example, the system would look for an entity named half_adder. Since no other information is provided, the names, types and modes of all of the signals provided in the component declaration must exactly match that in the entity declaration. Recall that if you use the **port map** and **configuration** statements, this will not be necessary.

8.5.3 Simulation Setup

Now that the model is loaded, we may want to generate test cases and provide stimulus to the model to determine if the model is indeed operating correctly. Generally, most simulators will provide for ways in which to specify a waveform on an input port of the entity being loaded. There will also be facilities for forcing signals to a certain value. Signal initialization, especially on input ports, is a necessary prelude to simulation. If we have constructed a model with a testbench format, then we may not need to do so. However, if we load a model such as the full adder into the simulator we will need to provide a stimulus to each of the inputs and examine the output signals to determine if the model is correct.

Example: Generating an Input Stimulus

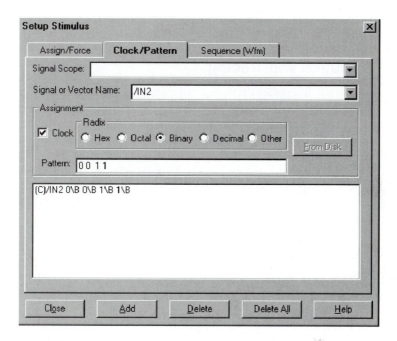

FIGURE 8-4 An example of creating stimulus on an input signal

Figure 8-4 shows an example of a dialog box from the Speedwave simulator in View-logic's Workview Office environment. A waveform is specified on the input signal IN2 . By denoting this waveform as a clock pattern, as simulation time progresses the waveform will be applied repeatedly until it is changed by the user. The notation is selected as binary. Signals may use hexadecimal notation, which is useful when applying a stimulus to 32-bit signals. Note the tab on the top of the box that states Assign/Force. This option is used to force a signal to a fixed value.

Example End: Generating an Input Stimulus

Another aspect of simulation is the notion of a simulator *step time*. Typically, you can advance simulation time in units called steps. For example, you might pick the step time as 10 ns. Running the simulation for 10 steps is analogous to running it for 100 ns. Simulators will provide facilities for setting the value of the step size. Step time can be important because often the input stimulus to a model being tested is applied for a period equal to the step time. For example, suppose you would like to create the following wave-form on an input signal named reset: 0 1 0 0. The signal reset should be driven to logic

0 for some duration and then logic 1 for some duration, and so on. For how long is `reset` driven to each value? The duration is generally equal to the simulator step time. If it is necessary to have a pulse of a specific width on `reset`, then the step time should be adjusted accordingly.

8.5.4 Execution and Tracing

We are finally ready to execute the model and trace the values of signals. Typically, execution can be initiated by a `run` or `step` command. The former starts a simulation for a fixed period of time or a fixed number of simulator steps. The latter steps through a single simulation cycle. To understand the timing behavior of our models, it is useful to be able to check the next set of events or signal value changes that are scheduled. This information is recorded in the event queue. Simulators generally provide facilities for examining the event queue and listing the scheduled events. Typically the values of all signals can be displayed in a trace window. *Remember: you cannot trace variables, only signals!* You can set break points and look at the value of the variables via simulator commands. However, variables do not exhibit time-dependent behavior in the same manner as signals and therefore cannot be traced in that fashion. Most simulators will also provide access to simulation statistics such as the number of events executed. This can provide useful insight into the behavior of the models as well as aid in debugging them.

Example: The Event Queue

In the Viewlogic Workview Office simulator, one can display scheduled events in the event queue. For the model shown in Figure 8-2 such a query might produce the information shown in Figure 8-5. The information provides the time at which the signals are to be assigned a new value. Note the way the signal is identified. In hierarchical models the path to a signal specified at some lower layer is specified in a manner that is very similar to the path names for files in directories.

```
show events
|At time = 110.0ns name = /H1/CARRY@/H1/process__on_line_4 Driver newvalue = '0'.
|At time = 110.0ns name = /H1/SUM@/H1/process__on_line_3 Driver newvalue = '1'.
|At time = 110.0ns name = /H2/CARRY@/H2/process__on_line_4 Driver newvalue = '0'.
|At time = 110.0ns name = /H2/SUM@/H2/process__on_line_3 Driver newvalue = '0'.
```

FIGURE 8-5 An example of the contents of the event queue

Example End: The Event Queue

Example: Signal Traces

All simulators provide some facility for tracing signals so that we may actively monitor the internal behavior of the model. An example of a trace window for our full-adder model is shown below. Note that initially, the internal signals and output signals of the model have undefined values. This is due to the fact that it takes some time, as determined by the gate delays of the components, for the input values to propagate to the internal signal paths and thence to the outputs.

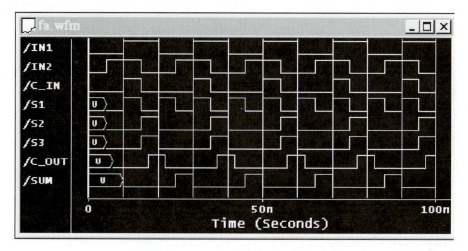

FIGURE 8-6 An example of a signal trace

Example End: Signal Traces

8.6 Chapter Summary

Traditionally, CAD tools have been large, complex aggregations of tools such as simulators, schematic capture tools, and layout editors. Although one can have a thorough knowledge of VHDL or VLSI design, and this knowledge may enable one to understand the concepts in using these tools, it also appears to be necessary to have some experience in the state of the practice to be able to use these tools successfully with minimal aggravation. This experience is often unrelated to design. This has often been referred to as "knowledge of the third kind." From a learning perspective, that situation is changing with the advent of low cost, relatively easy to use point tools such as VHDL simulators for PCs, Macs and workstation platforms. Since these are focused on VHDL, we can make use of them without having to navigate through the maze of complex CAD environments. However, there are still conventions, concepts, and state-of-the-practice notions that are acquired through experience. It is difficult to provide concrete actions, since these are dependent on the simulator toolset and its command set. However, we can introduce the common steps, concepts, and operations that will be encountered in the use of almost any simulator. Understanding how simulator environments are structured and how these structures are related to language concepts, is necessary for productive application to VHDL modeling. This chapter has attempted to provide an intuition about the practical aspects of VHDL simulator environments and hopefully ease the transition into proficient use. Often considerable frustration can be avoided by simply knowing what questions to ask and what to expect. This chapter, coupled with detailed tutorials for specific VHDL simulators provided in the Appendices, will hopefully bring the reader closer to that goal.

The concepts covered in this chapter include the following.

- Basic design units
  - entity
  - architecture
  - configuration
  - package declaration
  - package body
- Analyzing VHDL programs
  - order of analysis of design units
- Elaboration
  - elaboration of the design hierarchy
  - elaboration of the declarations
  - storage allocation
  - initialization
- Initialization of VHDL programs
- Simulation

CHAPTER 9 Identifiers, Data Types, and Operators

The VHDL language provides a rich array of data types and operators along with rules that govern their use. The goal of this chapter is to provide ready access to the syntax and semantics of the most commonly used data types and operators. This chapter is intended to serve more as a guide when writing your first VHDL programs rather than as a comprehensive language reference source. The more advanced language features are not referenced here but can be found in a variety of excellent texts on the VHDL language. Familiarity with common programming language concepts and idioms is assumed.

9.1 Identifiers

Identifiers are used as variable, signal, or constant names as well as names of design units such as entities, architectures, and packages. A basic identifier is a sequence of characters that may be upper or lower case, the digits 0–9, or the underscore ("_") character. The VHDL language is not case sensitive. The first character must be a letter and the last character must not be "_". Therefore `Select`, `ALU_in` and `Mem_data` are valid identifiers, while `12Select`, `_start`, and `out_` are not valid identifiers. Such identifiers are known as *basic identifiers*.

9.2 Data Objects

VHDL'93   In VHDL'87 there are three classes of objects: signals, variables, and constants. In
☞   VHDL'93 files are introduced as a fourth class of object. In VHDL'87 file types represent
          a subset of a variable object type. The range of values that can be assigned to a signal, vari-
          able, or constant object is determined by its type.

A signal is an object that holds the current and possibly future values of the object. In
keeping with our view of VHDL as a language used to describe hardware, signals are typically thought of as representing wires. They occur as inputs and outputs in port descriptions, as signals in structural descriptions, and as signals in architectures. The signal
declarations take the following form:

signal *signal_name*: *signal_type*:= *initial_value*;

Examples include

signal status : std_logic:= '0';

signal data : std_logic_vector (31 **downto** 0);

Recall that signals differ from variables in that signals are scheduled to receive values at some point in time by the simulator. Variables are assigned during execution of the
assignment statement. At any given time, multiple values may be scheduled at distinct
points in the future for a signal. In contrast, a variable can be assigned only one value at
any point in time. As a result, the implementation of signal objects must maintain a history
of values and therefore require more storage and exact higher execution time overhead
than variables.

Variables can be assigned a single value of a specific type. For example, an integer
variable can be assigned a value in a range that is implementation dependent. A variable of
type real can be assigned real numbers. Variables are essentially equivalent to their conventional programming language counterparts and are used for computations within procedures, functions, and processes. The declaration has the following form:

variable *variable_name*: *variable_type* := *initial_value*;

Examples include

variable address: **bit_vector**(15 **downto** 0) := x"0000";

variable Found, Done: **boolean** := FALSE;

variable index: **integer range** 0 **to** 10:=0;

The last declaration states that the variable index is an integer that is restricted to
values between 0 and 10 and is initialized to the value 0.

Constants must be declared and initialized at the start of the simulation and cannot be changed during the course of the simulation. Constants can be any valid VHDL type. The declaration has the following form.

constant *constant_name*: *constant_type* := *initial_value*;

Examples include the following.

constant `Gate_Delay`: **Time**:= 2 ns;

constant `Base_Address`: **integer**:= 100;

The first declaration states that the constant is of type **time**. This is a type unique to hardware description languages. Just as an integer variable can be assigned only integer values, the values assigned to the constant `Gate_Delay` must be of type time, such as 5 ns, 10 ms, or 3 s. In the above example, `Gate_delay` is initialized to 10 ns.

9.3 Data Types

The type of a signal, variable, or constant object specifies the range of values it may take and the set of operations that can be performed on it. The VHDL language supports a standard set of type definitions as well as enabling the definition of new types by the user.

9.3.1 The Standard Data Types

The standard type definitions are provided in the package STANDARD (see Appendix D.1) and include the types listed in Table 9-1. Note the definitions of **bit** and **bit_vector** types. From Chapter 2 we know that a simple 0/1 value system is not rich enough to describe the behavior of single-bit signals. This is why the community has moved towards standardization of a value system that multiple vendors can use. Such a standard is the IEEE 1164 value system, which is defined using enumerated types.

9.3.2 Enumerated Types

Although the standard types are useful for constructing a wide variety of models, they fall short in many situations. We know that single-bit signals may be in states that cannot be represented by 0/1 values. For example, signal values may be unknown, or signals may left

TABLE 9-1 Standard Data Types Provided within VHDL

Type	Range of values	Example declaration
integer	implementation defined	**signal** index: **integer**:= 0;
real	implementation defined	**variable** val: **real**:= 1.0;
boolean	(TRUE, FALSE)	**variable** test: **boolean**:=TRUE;
character	defined in package STANDARD	**variable** term: **character**:= '@';
bit	0, 1	**signal** In1: **bit**:= '0';
bit_vector	array with each element of type bit	**variable** PC: **bit_vector**(31 **downto** 0)
time	implementation defined	**variable** delay: **time**:= 25 **ns**;
string	array with each element of type character	**variable** name : **string**(1 **to** 10) := "model name";
natural	0 to the maximum integer value in the implementation	**variable** index: **natural**:= 0;
positive	1 to the maximum integer value in the implementation	**variable** index: **positive**:= 1;

floating. The language does support the definition of new language types by the programmer and the ability to provide functions for operating on data that is of this type. For example consider the following definition of a single bit:

```
type std_ulogic is ('U' -- uninitialized
                    'X' -- forcing unknown
                    '0' -- forcing 0
                    '1' -- forcing 1
                    'Z' -- high impedance
                    'W' -- weak unknown
                    'L' -- weak 0
                    'H' -- weak 1
                    '-' -- don't care
                    );
```

Now, assume we declare a signal to be of this type.

signal carry: std_ulogic := 'U';

The signal carry can now be assigned any one of the values defined above. Note that operations such as AND, OR, and "+/-", must be redefined for this data type. The type definitions and the associated operator and logical function definitions can be provided in a package that is referenced by your model. The above type definition is a standard defined by the IEEE and the associated package is referred to as the IEEE Standard Logic

1164 package (see Appendix D.3). This package is gaining popularity for two reasons. It provides a type definition for signals that is more realistic for real circuits. Furthermore, use of the same value system makes it easier for designers to share models, increasing interoperability, and consequently reducing model cost.

The above type definition is referred to as an *enumerated type*. The definition explicitly enumerates all possible values that a variable or signal of this type can assume. Another example of where enumerated types come in handy is the following:

type `instr_opcode` **is** ('add', 'sub', 'xor', 'nor', 'beq', 'lw', 'sw');

An instruction set simulation of a processor may have a large case statement with the following test:

case `opcode` **is**

when beq =>

Each branch of the **case** statement may call a procedure to simulate the execution of that particular instruction. Such type declarations can be made and placed in a package or in the declarative region of the **process**. For an example of the definition, declaration, and use of a type definition for memory in a simulation model of a simple processor, see the example code in Figure 4-2.

9.3.3 Array Types

Arrays of bit-valued signals are common in digital systems. An array is a group of elements, all of the same type. For example, a word is an array of bits and memory is an array of words. A common practice is to define groups of interesting digital objects as a new type. For example,

type `byte` **is array** (7 **downto** 0) **of bit**;

type `word` **is array** (31 **downto** 0) **of bit**;

type `memory` **is array** (0 **to** 4095) **of** `word`;

Now that we have created these new types, we can declare variables, signals, and constants to be of these types.

signal `program_counter` : `word`;= x"00000000";

variable `data_memory` : `memory`;

The use of type definitions in this manner enables us to define elements that we use when designing digital systems. For example, the preceding declarations demonstrate how we could define new types for words, registers, and memories. These new types make writing VHDL models more intuitive as well as easier to comprehend.

9.3.4 Physical Types

Physical types are motivated by the need to represent physical quantities such a time, voltage, or current. The values of a physical type are defined to be a measure such as seconds, volts, or amperes. The VHDL language provides one predefined physical type: **time**. The definition of the type **time** can be found in the package STANDARD. The definition from this package is reproduced below.

type time **is range** *<implementation dependent>*
units
fs;   -- femtoseconds
ps = 1000 fs;   -- picoseconds
ns = 1000 ps;   -- nanoseconds
us = 1000 ns;   -- microseconds
ms = 1000 us;   -- milliseconds
s = 1000 ms;   -- seconds
min = 60 s;   -- minutes
hr = 60 min;   -- hours
end units;

The first unit is referred to as the base unit and is the smallest unit of time. All of the other units can be defined in terms of any of the units defined earlier. For example, they all could have been defined in terms of femtoseconds. We can see how it is possible to define other physical types such as distance, power, or current. For example, we might define a type power as follows:

type power **is range** 1 to 1000000
units
uw;   -- base unit is microwatts
mw = 1000 uw;   -- milliwatts
w = 1000 mw;   -- watts
kw = 1000000 mw;   -- kilowatts
mw = 1000 kw;   -- megawatts
end units;

Note how kilowatts are defined in terms of milliwatts rather than watts. Now we can declare signals or variables to be of this type and assign or compute values of this type.

variable chip_power: power:= 120 mw;

The above declaration creates a variable of type power and whose value will be in the units defined above for variables of type power. When we are modeling physical systems it is very useful to have the ability to define physical types and the units that can be used to express their values.

9.4 Operators

Operators are used in expressions involving signal, variable, or constant object types. The following are the sets of operators as defined in the VHDL language [5]. Note that the shift **VHDL'93** operations are new in VHDL'93. The miscellaneous operators include **abs** for the computation of the absolute value, and ****** for exponentiation. The latter can be applied to any integer or real signal, variable, or constant.

logical operators	**and**	**or**	**nand**	**nor**	**xor**	**xnor**
relational operators	=	/=	<	<=	>	>=
shift operators	**sll**	**srl**	**sla**	**sra**	**rol**	**ror**
addition operators	+	–	**&**			
unary operators	+	–				
multiplying operators	*	/	**mod**	**rem**		
miscellaneous operators	**	**abs**	**not**			

The classes of operators shown are listed in increasing order of precedence from top to bottom. Operators with higher precedence are applied to their operands first. All of the operators within the same class are of the same precedence and are applied to operands in textual order—left to right. The use of parentheses can be used to define explicitly the order of precedence. In general, the operands of these operators must be of the same type, whereas the type of the permissible operands may be limited. The following tables provide information from the VHDL Reference Manual [5] and define the permissible operand types.

TABLE 9-2 Operator–operand relationships

Operator	Operand Type	Result Type
=	Any type	Boolean
/=	Any type	Boolean
<, >, <=, >=	Scalar or discrete array types	Boolean

VHDL'93

The shift operators are defined in Table 9-3. These operators are new in VHDL'93, and are particularly useful in describing operations in models of computer architecture components. Some examples of the application of these shift operators are provided in Table 9-4.

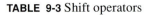

TABLE 9-3 Shift operators

Operator	Operation	Left operand type	Right operand type	Result type
sll	Logical left shift	Any one dimensional array type whose element type is **bit** or **Boolean**	**integer**	left operand type
srl	Logical right shift	"	"	"
sla	Arithmetic left shift	"	"	"
sra	Arithmetic right shift	"	"	"
rol	Rotate left logical	"	"	"
ror	Rotate right logical	"	"	"

TABLE 9-4 Examples of the application of the shift operators

Example expression	Result operand value
value <= "10010011" sll 2	"01001100"
value <= "10010011 sra 2	"11100100"
value <= "10010011" ror-3 (i.e., rotate left)	"10011100"
value <= "10010011" srl-2	"01001100"

The addition, subtraction, and concatenation operators are described in Table 9-5. The first two are self explanatory.

TABLE 9-5 Addition and subtraction operators

Operator	Operation	Left operand type	Right operand type	Result type
+ or -	Addition/subtraction	Numeric type	Same type	Same type
&	Concatenation	Array or element type	Array or element type	Same array type

The concatenation operator composes operands. For example, we might have:

result(31 **downto** 0) <= '0000' **&** jump (27 **downto** 0);

The upper four bits of result will be cleared and the remaining 28 bits will be set to the value of the 28 least significant bits of jump.

The unary operators are described in Table 9-6.

TABLE 9-6 Unary operators

Operator	Operation	Operand type	Result type
+/-	Identity/negation	Numeric type	Same type

Finally, some remaining numerical operators are illustrated in Table 9-7.

TABLE 9-7 Numerical operators

Operator	Operation	Left operand type	Right operand type	Result type
* or /	Multiplication or division	Integer or floating point type	Same type	Same type
mod or rem	Modulus/remainder	Integer type	Same type	Same type

9.5 Chapter Summary

This chapter has provided a brief overview of the common objects, types, and operators in the VHDL language. I have taken the approach of focusing on the unique aspects of the VHDL language throughout most of the text while assuming that language features such as types, identifiers, and objects, are familiar concepts to the reader and a handy reference is all that suffices as a prelude to writing useful models. This chapter is intended to fill that role and will be only as useful as the previous chapters have been successful in providing an intuitive way of thinking about, constructing, and using VHDL models. Familiarity at this level can lead to the next level in using the more powerful (and complex) features of the language.

References

1. J. Bhaskar, *A VHDL Primer*. Englewood Cliffs, NJ: Prentice Hall, 1995.

2. B. Cohen, *VHDL Coding Styles and Methodologies*. Boston, MA: Kluwer Academic, 1995.

3. D. Gajski and R. H. Kuhn, "Guest Editors Introduction—New VLSI Tools," *IEEE Computer*, vol. 16, no.2, 1983, pp. 14–17.

4. J. Hayes, *Introduction to Digital Logic*. Reading, MA: Addison-Wesley, 1993

5. *IEEE Standard VHDL Language Reference Manual: ANSI/IEEE Std 1076–1993*. New York : IEEE, June 1994.

6. R. Lipsett, C. Schaefer, and C. Ussery, *VHDL: Hardware Description and Design*. Boston : Kluwer Academic, 1989.

7. V. K. Madisetti, "Rapid Digital System Prototyping: Current Practice and Future Challenges," *IEEE Design and Test*, Fall 1996, pp. 12–22.

8. V. K. Madisetti and T. W. Egolf, "Virtual Prototyping of Embedded Microcontroller–Based DSP Systems," *IEEE Micro*, pp. 9–21, 1995.

9. D. Patterson and J. Hennessey, *Computer Organization & Design: The Hardware/Software Interface*. San Francisco : Morgan Kaufmann, 1994.

10. D. Perry, *VHDL*. New York, NY: McGraw Hill, Second Edition, 1994.

11. M. Richards, "The Rapid Prototyping of Application–Specific Signal Processors Program," *Proceedings of the First Annual RASSP Conference*, Defense Advanced Research Projects Agency, 1994.

12. K. Skahill, *VHDL for Programmable Logic*. Reading, MA: Addison-Wesley, 1996.

13. D. E. Thomas, C. Y. Hitchcock III, T. J. Kowalski, J. V. Rajan, and R. A. Walker, "Automatic Data Path Synthesis," *IEEE Computer*, vol. 16, no. 12, December 1983, pp. 59–70.

14. R. Walker and D. E. Thomas, "A Model of Design Representation and Synthesis," *Proceedings of the 22nd ACM/IEEE Design Automation Conference*, 1985 pp. 453–459.

Simulator Tutorial: Workview Office

This tutorial steps the user through the process of analyzing and simulating a simple gate-level model using the Workview Office 7.11 toolset. Commonly performed operations are described—analyzing the models, viewing the event list, generating stimulus, and so on. Most commercial simulators will support similar features but may differ in the layout and organization of the menus. The examples in this text were tested with this toolset. However, some of the examples use the `std_logic_arith` package. This library was compiled into the library `WORK`. Other simulators will have similar packages possibly in system libraries (for example, possibly in library `IEEE`) but may have similar functions with different names. We have endeavored to keep everything else within the example compliant with VHDL'87.

Although any one of the models described in this text may be used in this tutorial, we recommend that the user start with a relatively simple model, such as a behavioral model of a half-adder, full adder, or multiplexor. Throughout this tutorial we will assume that the reader is using the VHDL model of a full adder shown in Figure 3-3 as a test case. Setting up and simulating a model consists of three basic steps: (i) setting up a project, creating a project design library, and selecting any other libraries to be used during simulation, (ii) analyzing the VHDL source files with the `Analyzer`, and (iii) simulating the model using `Speedwave`, in the Workview Office suite.

Step 1: Getting Started

Workview Office tools require a working directory where all the files related to a particular project are kept. This is referred to as the *project design directory*. Place your VHDL source files in this project directory and when they are analyzed, the resulting executable files and other files used by the simulator are stored in this directory. The project design directory is created with the `Project Manager` tool.

- First, create a directory such as *a:\project,* to hold your VHDL source files. Creating your design directory on a floppy will be somewhat slower than creating it on the internal hard drive. However, it keeps the design directory portable across any of the machines in a laboratory. If you have a personal machine, it is better to create the design directory on the hard drive.

- Invoke the `Project Manager` from Workview Office. Select `browse` and select the project directory that you just created. Ensure this directory appears in the `browse` dialog box. Accept this directory and control now returns to the `Project Manager`.

- From the `File` menu item, select `Save As`. The `Save As` dialog box will prompt you for the project name. You can accept the default name (usually this is *user.vpj*). After the save operation is complete control returns to the `Project Manager`.

You have now created a project design directory and the `Project Manager` window will now appear as shown in Figure A-1. Note the primary directory is set to *a:\project*. This is now the default design directory for subsequent use by the Workview Office tools unless it is changed. Therefore in a laboratory setting before using any of the tools you should check and ensure that the project design directory is set to where you need it. If you wish to use a different project directory you must first set this in the `Project Manager` before invoking the `Speedwave` simulator.

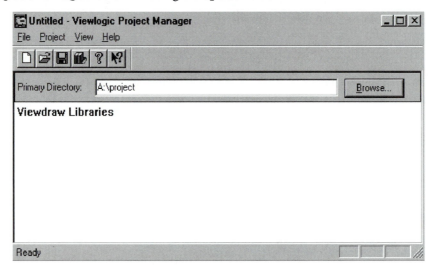

FIGURE A-1 Setting up the project

Step 2: Setting Up the Design Environment

Prior to compilation and simulation, you must set up the design environment. This includes libraries, paths, and the working directory. The concept of libraries is central to the organization and use of VHDL models. Your VHDL files are organized in your *project design library*. There are also system design libraries that you may use, such as IEEE, and you may declare these in your VHDL source files. This is akin to the declaration and use of program libraries in C or Pascal. When you start analyzing VHDL files and begin simulations, unresolved references to design units will be searched for in libraries that you declare. You need to specify all of the libraries that are to be searched and the order in which they will be searched. This list is placed in a file called *vsslib.ini* that is stored in the project directory. This step of the tutorial creates this file.

- Invoke the Speedwave design tool from Workview Office. From the Tools menu select VHDL Manager. You will initially get a message stating that there is no *vsslib.ini* file. You can ignore this message, as this is the file you are about to create.

- In the VHDL Manager window from the File menu select Create Library. From the resulting dialog box, select browse and subsequently select the directory that is your project directory. Your project design library now will be created. The default name for this library will appear as *user.lib*. Accept the default and select Save, causing the creation of *user.lib*.

- You will now see a Create Library dialog box with *A:\project\user.lib* appearing as the library path name. In this dialog box, provide a symbolic name for this library. A good suggestion is to provide the same symbolic name, in this case USER. Type USER and click on OK. The symbolic name can be used to reference the project design library in the same way that we use IEEE in all of the VHDL models in this text. What we have done is create your design library. This library has a name (*user.lib*) and can be referred to by its symbolic name (USER). Control returns to the VHDL Manager.

- Your user library, *user.lib*, appears in the VHDL Manager window; select it and click on the Add Lib button. If *user.lib* does not appear, click the List User Libs button and navigate the directories to find it. The *user.lib* library now appears in the lower half of the VHDL Manager window under Project Libraries.

- Similarly, you can click on the List Sys Libs button and select and add any system libraries that you need. Add *IEEE.lib*. You may find it useful to add other libraries, such as *pack1076.lib* or *math.lib*. These libraries also now appear under project libraries. **When you are done, make sure that your project design directory is at the top!** You can ensure this by selecting your project design directory from this list and clicking on Set Working. The reason you want your directory to be the first is that when the analyzer–simulator is looking for models and attempting to resolve references, this is the order in which the directories are searched.

- Now, select Save vsslib.ini from the File menu. Paths to all selected libraries are now stored in the *vsslib.ini* file in your project directory.

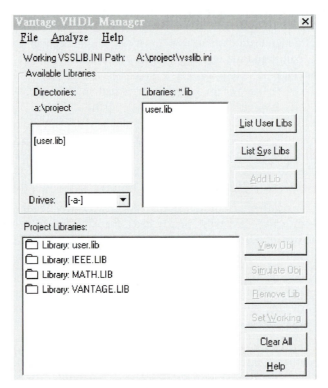

FIGURE A-2 The VHDL Manager window

You have now completed the project initiation component of using Workview Office. The VHDL Manager should now appear as shown in Figure A-2.

Step 3: Create the VHDL File

Use your favorite text editor to type in the sample VHDL file. Let us use the full-adder model from Figure 3-3. Do not use a word processor even though there may be an option for saving your text as an ASCII file. Some word processors place control characters in the file or may handle some characters non-uniformly, such as left and right quotation marks. This can lead to analyzer errors (which you could correct at that time). In this tutorial make sure that the entity and architecture descriptions are in the same file. VHDL is not case sensitive nor is it sensitive to white spaces which can be inserted to suit personal preferences. Call this file *first.vhd* and make sure it is in *A:\project*.

Step 4: Analyzing the VHDL File

In the Speedwave window under the Tools menu item click on the Analyze VHDL item. The Analyzer window pops up and will appear as shown in Figure A-3. Click the Settings button. Make sure that the appropriate box is checked to generate a listing file. If there are errors during compilation it is convenient to look at the listing file named *first.lis* to study the errors.

- Select the file *first.vhd* in the Analyzer window and click on Analyze.

- A separate window will pop up reporting on progress with the compilation of your program. If there are no errors you can close the Analyzer window and proceed.

- If there are errors reported, use a text editor to examine the file *first.lis* in your project directory *A:\project*. Find and correct any syntax errors (these are the only errors that your should have). Remember that one error can lead to many errors becoming reported on subsequent lines. Therefore, pay particular attention to the first reported error.

    The analyzer will create several new files in your project directory that are required by the simulator.

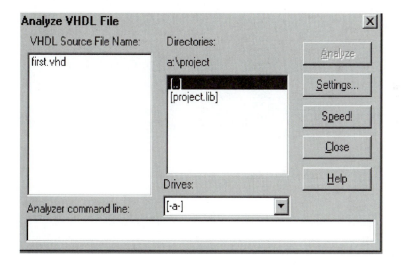

FIGURE A-3 Analyzing VHDL files

Step 5: Loading and Simulating the VHDL Model

We are now ready to load the compiled model into the simulator.

- In the `Speedwave` window under the `File` menu item, select `Load VHDL Design`. A dialog box will pop up with your design library, as shown in Figure A-4.

- Click on the design library *user.lib* and you will find the names of design entities contained in the design library. In this case you will only have the design entity `full_adder`. Note that this is the name in the entity declaration and not the file name. As far as the toolset is concerned, it is the name of the design entities that matter.

- Double click on the entity name to load it into the simulator. The model is now loaded. Alternatively, click on the entity name once to generate the command that appears on the command line and then click `OK`.

After the model has been loaded, the `Speedwave` window will appear as shown in Figure A-5. You will notice that two windows have been created. One contains the VHDL source for the model. The second is the `Navigator`. This tool enables you to traverse the design and select signals in a hierarchical design in a graphical manner. For example, one of the lines on the right side of the `Navigator` should indicate the number of signals in the model (e.g., 8 for the full adder). If you click on this icon, the list expands to name all

FIGURE A-4 Selecting the design entity for simulation

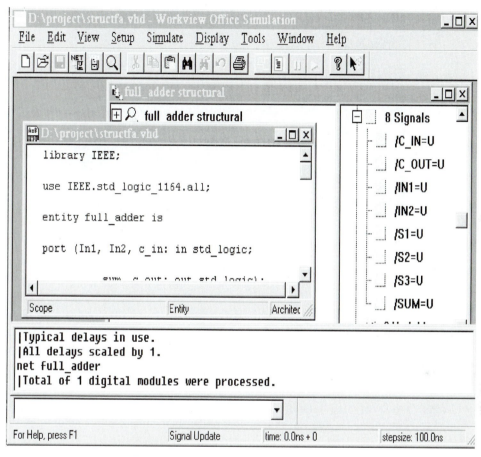

FIGURE A-5 Speedwave window

of the signals as shown in the figure. Keep this expanded list in the Navigator window for a later step.

Step 6: Setting the Simulation Time Step

Simulation proceeds as a sequence of "steps". Each step is of a specific time duration. The user deals with the simulator in terms of steps. For example, you may instruct the simulator to simulate for 10 time steps. How long is 10 time steps? The answer is: as long you want it to be. You can set the duration of a step as follows.

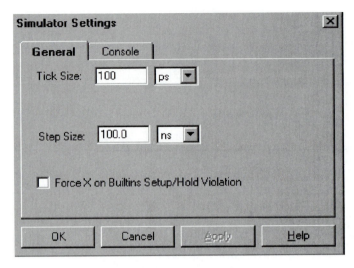

FIGURE A-6 Simulator settings

- Under the Setup menu click on Simulator Settings. In the resulting dialog box, shown in Figure A-6, you can set the step size. Set this time to 10 ns. Note that within the bottom half of the Speedwave environment the following command appears:

    ```
    stepsize 10ns
    ```

    Make a note of this command. Later we will see how these commands can be retained in a command file that can be executed rather than going through all of these dialog boxes.

Step 7: Generating Input Stimulus

Now that the model has been loaded we must provide a stimulus to the input signals. For the full-adder model that we are working with, we should provide a set of waveforms on the input ports. Alternatively, you can force individual signals to 0 or 1 values.

- Select the Stimulus item under the Setup menu item in the Speedwave window. This will produce the dialog box shown in Figure A-7. At the top of the dialog box you can select a tab for Assign/Force or Clock/Pattern. The former will enable you to assign a specific value to an input signal. In this example, let us select the latter.

- Select the signal IN2 from the Navigator window. After selecting the signal, you must select the dialog box for the signal name to appear in the window as shown in Figure A-7. Alternatively, you can type in the signal name.

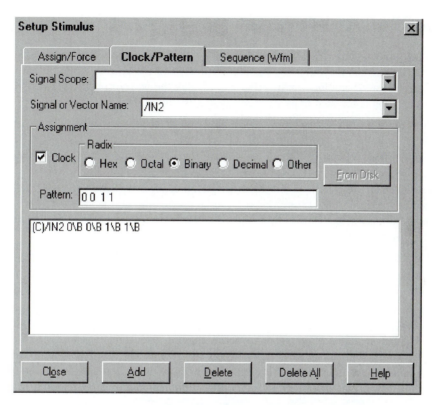

FIGURE A-7 Application of a stimulus to the input signals

- In the dialog box, select clock and binary values. In the pattern window, type a pattern of signal values that you wish the signal to have. Make sure there are spaces between the values. By denoting the type clock, these signal values are applied repeatedly. Click the Add button. Repeat for all three inputs. When the Add button is clicked, note the command in the bottom portion of the Speedwave window. We will use these in command files later. If we had 32-bit quantities as input, we could have specified the values in hexadecimal notation. Make sure all inputs have a value, either a waveform input or a forced value of 0 or 1.

The preceding sequence of operations sets up the simulator to apply the given sequence of signal values to the input signals In1, In2, and C_in. For example, the value 0 will be applied to the In2 input for a period of one step time (the duration of a step was set in Step 6), followed by a value of 0 for the next step time followed by value of 1 for the next two step times.

Step 8: Setting Up Streams

When the model is simulated we would naturally like to view the waveforms generated on the signals. To trace the values of signals, we must first set up a connection between the signals to be traced and the window that will display the waveforms of the traced signals. This connection is referred to as a *stream*, as in a stream of data.

- From the `Speedwave` window select `Setup` and then `Waveform Streams`. The resulting dialog box will appear as shown in Figure A-8. Select `New Stream`. You will be asked to supply a filename. The extension will be automatically set to denote a file containing waveform data (*.wfm*). Use the name *first* so the waveform file will be *first.wfm*. This file will contain the descriptions of the waveforms of the signals being traced. Note that in the future this file can be opened directly from the Trace Window. Now you must select signals that you would like to trace, and record this information. This can be done as follows.

- Select a signal from the `Navigator` window. Then click on `Add Signal` in the `Setup Waveform Streams` dialog window. This will add a signal to the list of signals to be traced. Add all 8 signals in the full adder model.

- When you have selected all of the signals that you want to trace, click the `Apply` button. This will create the waveform file and start up the `ViewTrace` window.

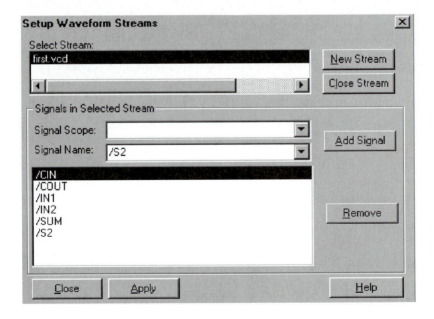

FIGURE A-8 Setting up a waveform stream

If you have an existing waveform file, then select it from the list. The list of signals being traced in this file is shown below. You can add to, or delete from, this list. An existing file can be opened directly from the `ViewTrace` window.

Step 9: Simulation and ViewTrace

At this time you should have a `ViewTrace` window open, graphically illustrating all of the signals that you have selected to be traced. Every time the simulation proceeds through a step, this display will be updated to show the changes in the values of the signals being displayed. After the simulation has progressed through a few steps, the trace window will appear as shown in Figure A-9.

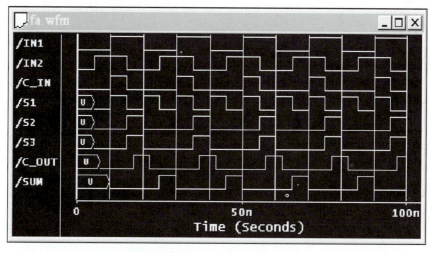

FIGURE A-9 A trace window

Step 10: Running the Simulation

We are finally ready to actually run a simulation. The duration of the simulation can be expressed in absolute units, such as nanoseconds, or in terms of simulator step times. Recall that the duration of a step had been set earlier.

- Select `Run` under the `Simulate` menu item. From the dialog box set the specified run time to 100 ns. Click `Apply` and observe the trace window. Note that the input waveform on `In1` is applied repeatedly over time.

- Experiment and look at the results in the `ViewTrace` window. You can go back and change the input waveforms on the input signals and continue to simulate. Check the displayed traces and make sure that the model is correct. Be careful! The gate delays may present a waveform different from the one you might expect to find in a textbook! Cross-hatched displays indicate that the signal value is undefined. Experiment with the F6, F7, and F8 function keys. If the cursor is positioned in the left (right) half of the trace, F6 will pan to the left (right). The F7 and F8 keys will enable zoom into and out of the trace.

- Examine the relative delays between the values of the output and the corresponding input values.

- Click on the `Display` button on the top menu bar. You can examine simulation statistics and list pending events. Experiment. You can see the next set of signal transitions that are scheduled and their scheduled values. Note the time associated with these transitions.

- Change the step size to different values. For a fixed number of simulation steps, distinct step times produce different simulation durations. The trace may appear a bit different (it actually just looks longer or shorter).

- Note that you can single-step and set breakpoints here. Signal values at a breakpoint can be examined via the `Display` button.

- Sometimes it is useful to be able to place the simulator console somewhere else. You can undock the console from the simulator window by selecting `View=>Dock`. This is a toggle switch.

You can annotate the trace by selecting `Annotation` from the `ViewTrace` menu and selecting `Add`. You can now type in a character string in the dialog box and place it anywhere on the trace. Experiment with the various options in the trace window.

Step 11: Command Files

Rather than always going through the same sequence of commands, you can place commonly used commands in a command file. Most operations that you perform via a dialog box can be executed from the command line at the bottom of the Speedwave window. Note the commands as you go through the dialog boxes. For example, you can open a trace window by executing the following command:

```
wave first.wfm  /In1  /In2
```

This command starts the trace facility with the waveform file *first.wfm* and traces the signals `In1` and `In2`. You can add signals to the trace during a simulation by executing this command in the command window at the bottom.

You can execute a command file by clicking on the `File` button on the top menu bar in the `Speedwave` window and then selecting `Run Command File`. The simulator will execute the set of commands in the file that you specify. This feature is particularly useful when you are debugging and rerunning simulations repeatedly. You can determine what commands to place in the file by observing the command that shows up in the simulator console window whenever you execute some operation in the simulator. You can set it up so that the command file sets up the stimulus, records signals to trace, sets the simulator step time, opens the trace window and runs the simulation for 500 ns. An example of such a command file is provided for the VHDL model of a datapath described in Appendix C.

Step 12: Plotting Trace Analyzer Output

- Use `Print Preview` to check the output trace before printing.
- The trace can be printed from the `Print Preview`.

Step 13: Miscellaneous Useful Features

There are several other features of the toolset that are not discussed here and can be readily explored.

- Use the help facility to learn more about various commands available in the `Speedwave` environment.
- Look through the menus in the `Speedwave` and `ViewTrace` utilities and experiment with them to try and understand what they do.
- Under `Setup` you can set the radix of the signals for display purposes by selecting `Vector Radix`, and a signal from the `Navigator` Window. This is particularly useful if you have 32-bit quantities. It is usually preferable to display such quantities in hexadecimal notation. Otherwise the signal will be displayed as a 32-bit number. This can take up a great deal of space in the trace window.
- Sometimes the value of a signal will not be displayed in the trace. To see the value, you may have to zoom in using the F7/F8 keys. By zooming in and making the visible area of the trace larger, you will eventually see values associated with the signals.
- Although the trace window display is in color, most printing is not performed in color. Some colors reproduce better than the others. Within the trace window, select a signal and change its color by selecting `Properties` under the `Signals` menu item.

Step 14: Exit

This tutorial has exercised only a subset of the features of the Workview Office tools. Although it has covered the basic operations in the creation, analysis, and simulation of VHDL programs the reader should be encouraged to explore the remaining capabilities of the toolset.

From the main window, click on the `File` button and select `Exit` from the menu.

Simulator Tutorial: Cypress WARP

This tutorial steps the user through the process of analysis and simulation using the WARP2, release 4.1 toolset from Cypress Semiconductor. This toolset is quite different in the approach used to simulate a VHDL design. This text is focused on the use of VHDL as a simulation language. We model the behavior of digital systems in terms of the generation of events (signal value transitions) and the times at which these events occur. Such a view enables us to construct high-level models based on behavior without having a specific hardware implementation. In contrast, the use of the WARP toolset is predicated on the *synthesis* of a hardware design corresponding to the VHDL model and the subsequent simulation of this hardware implementation. Although this is a different view of the use of VHDL, the inexpensive availability of tools such as WARP make it an attractive starting point for learning basic VHDL modeling. The starter models we have used here should not present any problems for the WARP synthesis compiler, which is focused on synthesis for the low cost, flexible end of the market using programmable logic devices. However, as we proceed to more complex models, a more thorough understanding of the synthesis process is necessary to create VHDL models that can be synthesized to produce efficient solutions. At this juncture we may also be interested in more powerful (and more expensive) synthesis tools capable of accommodating larger, more complex designs for implementations using programmable logic devices or higher performance application specific custom solutions.

In the past decade there has been an explosion of work in the area of the synthesis of digital logic: the generation of hardware designs from high-level language specifications. Hardware description languages such as VHDL and Verilog serve as the specification medium for the system behavior. Synthesis tools such as WARP generate hardware designs to implement the functionality of these specifications. Before we discuss the use of the WARP toolset, it is instructive to briefly distinguish between simulation and synthesis. This distinction is important, since it affects the types of models that one can study with the WARP toolset.

Setting up and simulating a model consists of three basic steps: (i) setting up a project and creating a project design library, (ii) analyzing the VHDL source files and synthesizing the hardware implementation with `Galaxy`, and (iii) simulating the model using `Nova`, the WARP simulator. Synthesis is an active area of industrial and academic research, and has been for some time now. The use of VHDL for synthesis is a topic worthy of a text and has been effectively treated in other texts (e.g., [10] and [12]). My goal in the following section is to highlight some of the differences between simulation and synthesis so that we might use the WARP tool effectively for our purpose, namely getting a quick start on the use of the VHDL language.

B.1 Simulation vs. Synthesis

Suppose you were asked to design a digital circuit to perform a specific task. You would need some description of the computation to be performed by this circuit. In general, this description could take many forms, for example, state diagrams, logic diagrams, or truth tables. We are interested in descriptions presented in VHDL. How would you proceed to generate a hardware design from such a description? You would first need to know the hardware primitives that were available for you to work with. Do we have decoders and multiplexors? What type of flip-flops can we use? How about the types of logic gates that are available? With answers to these questions we would attempt to realize the design using the available primitives.

We have already had some experience with elementary synthesis in our classes in digital logic. From Boolean equations, truth tables, and state diagrams we follow well-defined procedures for generating hardware designs. When designs become too complex, we rely on computer aided design tools to help us generate the designs. However, the design specification for such tools is usually at quite a low level such as truth tables or state diagrams. Higher-level synthesis enables us to describe our designs in VHDL and synthesize to a target set of hardware primitives. Since these primitives are known hardware components (e.g., gates, multiplexors, registers), accurate timing information is available and very accurate simulation models can be constructed. The WARP simulator falls into this category of VHDL tools. The user describes the behavior of the circuit to be simulated in VHDL. After selecting a target library of primitive components the WARP compiler creates a hardware design from these primitives to perform the functions captured in the VHDL model. This hardware design can be simulated using the WARP tools, or simulation models can be created for VHDL simulators available from vendors such as Viewlogic and Model Technologies.

Why synthesis? Imagine that we could describe at a very high level the operation of a complex modern microprocessor and have software tools synthesize an efficient design.

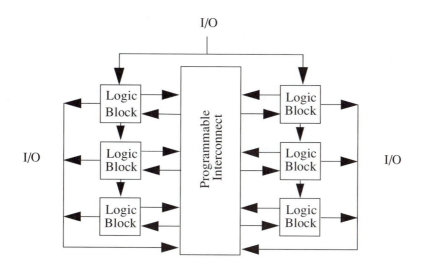

FIGURE B-1 Example of the architecture of a programmable logic device

We could dramatically reduce the time needed to design and bring the product to market not to mention the enormous savings in development cost. Leading research efforts are constantly pushing the limits on the size and complexity of circuits that can be synthesized and the parameters that can be optimized during this process. Examples of optimization criteria include chip area, speed, and power. We may not expect a synthesized design to perform as well as a custom design, but the cost–performance ratio may be superior. For many low-end designs that tend to be cost-sensitive, synthesis may be the preferred approach.

Popular classes of configurable hardware devices include programmable logic devices (PLDs) and field programmable gate arrays (FPGAs). These devices are populated with blocks of logic that can be programmed to implement various Boolean functions. The interconnect between the logic blocks and the inputs/outputs are similarly programmable. Such configurable hardware is becoming increasingly popular for low-cost applications as well as for those applications that require rapid reconfiguration to suit new requirements. An example of the architecture of such a device is shown in Figure B-1. Configurable logic blocks can be programmed to implement logic functions while the interconnect can be programmed to provide connections between the logic blocks and I/O pins. The WARP synthesis compiler produces a hardware description of the logic circuit that implements the VHDL function, which is then *placed* and *fitted* into the PLD. The result is a file that can be used to program such a PLD device.

The WARP tools can create a VHDL model of the synthesized and fitted design. As we might imagine, such as design is probably quite accurate if we draw upon a library of hardware primitives (for which accurate timing information is available) for synthesis. The VHDL model can be used for simulation using any one of a number of commercially available simulators. Alternatively, we may use a functional simulator available within WARP to simulate the hardware description using the file that is used for programming the PLD.

Note that this process provides new capabilities as well as restrictions compared to what we can do with a simulator such as Speedwave in Workview Office as described in Appendix A. Discrete event simulation is quite general and we can simulate the behavior of computer architectures at a very high level. Such simulations enable architectural trade-offs to be made early in the design process, when it is relatively easy to make changes. However, high-level models are harder to synthesize and high-level synthesis is currently the subject of intense research focus. To make the synthesis process tractable, most compilers constrain the VHDL models for synthesis by placing restrictions on the VHDL constructs that can be used as well as placing restrictions on programming style. Thus, we should be aware that currently only a subset of VHDL programs will be successfully synthesized. We should also be aware of limitations that are often placed on use of the VHDL constructs within these programs. The goal of this appendix is not the in-depth discussion of issues of synthesis vs. simulation. Rather for early exposure to the use of VHDL the WARP tools provide an inexpensive option. We can learn the basic VHDL constructs from the application to models of digital logic circuits. This appendix is intended to aid in its utility in this context. Given this background, we are ready to proceed with the tutorial on the use of the WARP tools.

B.2 Using the WARP Tools

Step 1: Getting Started

Like most simulators, the WARP tools use the concept of a project. When you invoke the Galaxy synthesis compiler, if a project does not exist, you will be requested to provide a design directory and project name to be created. A file with the extension *.wpr* will be created. Alternatively, you can create a new project from the Galaxy window by selecting New under the Project option. The project file contains information about the environment and options for compiling a VHDL model. Create the project warp in a directory called *class*. You can do this by typing *C:\class\warp* when you first invoke Galaxy. In the directory *C:\class* you should then find the file *warp.wpr*, which contains information about the default working directory, library search order, and related information about the design environment. This will be the default design directory for use by the WARP tools. In the future when Galaxy is invoked, the default directory will be set to *C:\class* unless changed. If you wish to use a different project directory you must select Open from the Project menu item to open a new project design directory.

Step 2: Create the VHDL Files

Using your favorite text editor, create three VHDL files for the structural model of a full adder shown in Figure 5-2. One file contains the top-level structural model as shown in Figure 5-2. Name this file *structfa.vhd*. Two other files that contain the behavioral model of a half adder (*ha.vhd*) and a two-input OR gate (*or2.vhd*) must be constructed (see Chapter 3) and included in the project design directory. Do not use a word processor even though there may be an option for saving your text as an ASCII file. Some word processors place control characters in the file or may handle some characters non-uniformly, such as left and right quotation marks. This can lead to compiler errors (which you could correct at that time). In this tutorial, make sure that the entity and architecture descriptions for each design component are in the same file. The VHDL analyzer is not case-sensitive nor is it sensitive to white spaces which can be inserted to suit personal preferences.

Since we are using a synthesis compiler, specification of the propagation delays through the circuit is not warranted. The actual delays through the circuit will be determined by the hardware primitives that are used to realize the circuit, which in this case is a PLD implementation. **Therefore, do not specify delays in the VHDL models used in this tutorial!** The model can still be simulated in WARP and other toolsets, since we know from Chapter 3 that delta delays are used to ensure correct functional simulation in the absence of the specification of any delays.

Step 3: Add Files to the Current Project

After these files have been created make sure you place them in the project directory. Although they may be in the project directory, they are not a part of our current working project. The files in the current working project must appear in the Galaxy window. We can now add them to the current project from the Galaxy window by selecting Add under the Files option. You should now see a dialog box as shown in Figure B-2. Select each file and click on the right arrow to include this file in the project. When you have completed this operation, select OK and you will return to the Galaxy window, which should now appear as shown in Figure B-3.

The set of files in the current project are shown in the left part of the Galaxy window, and the order of the files is important. The files are ordered from top to bottom according to their position in the design hierarchy. Therefore, the last file in the list corresponds to the top-level file. This ordering does follow intuition since files lower on the list may use components defined in files higher on the list. Therefore, the listing reflects the order in which the files must be compiled. Only the top-level file is actually synthesized and, therefore, only one top-level file is permitted in a current project. Files can be added or deleted from a project window via the Files menu item. If all of the files in the project directory are to be included in the project, we can do this by simply selecting Add all from the Files menu. You can also rearrange the order of files through the same menu. The goal is to set the correct order of files in the project menu and to select one file as the top-level file. This is achieved by selecting the file *structfa.vhd* and clicking the Set

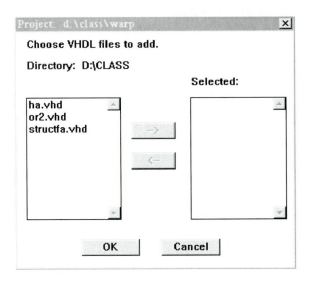

FIGURE B-2 Selecting files to include in a project

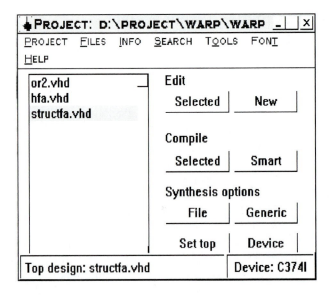

FIGURE B-3 The Galaxy window

Top button. We have now completed the setup of our project and can proceed to the next step: compilation and synthesis.

Step 4: Set Synthesis Options

There are four options under the Synthesis Options portion of the Galaxy window. Before we can begin the process of synthesis, we must select the target device that will host the synthesized hardware design. From the Synthesis Options menu, select the Device button. This will produce the dialog box shown in Figure B-4.

In the top left corner we can select the device that is the target of the synthesis operation. In this example, we select C371, a 32-macrocell complex programmable logic device (CPLD) from Cypress Semiconductor. The corresponding VHDL package is automatically selected, as shown when you select the device. This default package is usually the one that contains models corresponding to the highest-speed parts available. In the output portion of the dialog box we must select the form of the output. We select a format as a JEDEC file and select the form of the post-synthesis simulation. In this case we select a format compatible with the IEEE 1164 standard. The JEDEC file can be simulated with Nova: a simulator packaged with the WARP tools. From the list of options, you can also

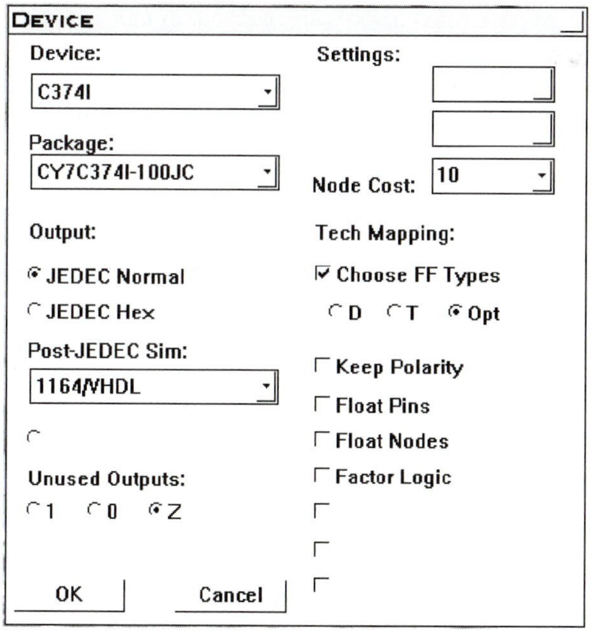

DEVICE

Device:
C374I

Package:
CY7C374I-100JC

Output:
● JEDEC Normal
○ JEDEC Hex

Post-JEDEC Sim:
1164/VHDL
○

Unused Outputs:
○ 1   ○ 0   ● Z

OK   Cancel

Settings:

Node Cost: 10

Tech Mapping:
☑ Choose FF Types
○ D   ○ T   ● Opt

☐ Keep Polarity
☐ Float Pins
☐ Float Nodes
☐ Factor Logic
☐
☐
☐

FIGURE B-4 Device selection dialog box

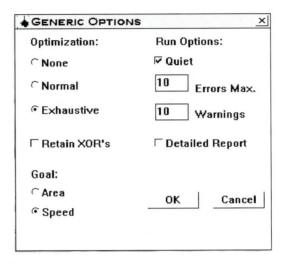

FIGURE B-5 Generic options

select output formats compatible with simulators available from other CAD vendors. Finally, we can select how unused outputs are handled. Retain the defaults for the remaining parameters.

Now that we have selected the target device and some of the synthesis options, let us select some optimization parameters. We can do this by selecting the `Generic` button from the `Galaxy` window. This will pop up the dialog box shown in Figure B-5. Generic optimization parameters can be set within this dialog box. Retain the defaults. The meaning of the parameters is largely self-explanatory. You can seek to optimize the speed of the resulting design at the expense of increased area or number of logic cells required. The level of optimization can also be selected, including retention of the XOR structure, which will produce better results for target devices that have XOR gates in the cell structure. Finally, you can also select the level of detail in the generation of messages during compilation and synthesis.

The `Set Top` button in the `Galaxy` window establishes the top-level file in the hierarchy. In our case, it is the file *structfa.vhd*. Select this filename in the `Galaxy` window and click the `Set Top` button. A message in the lower left portion of the `Galaxy` window will now indicate that the top-level design has been selected. There can only be one top-level design and this should be the last design in the list. The `File` button simply identifies the properties of a file that has been selected in the `Galaxy` window. For example, select the file *ha.vhd* and then click on the `File` button. The resulting dialog box will indicate that the file is to be compiled into the library WORK. Now, close this dialog box and select the file *structfa.vhd* in the `Galaxy` window. The `File` button dialog box will indicate that this file is the top level of the design.

Step 5: Edit Options

The Edit options bring up the editor to enable you to create new files or edit an existing VHDL file. For example, select the file *ha.vhd* and click on the Selected button. Alternatively, you may double click on *ha.vhd*. The editor will pop up with the contents of the file *ha.vhd*. The button New clearly is for creating new files.

Step 6: Compile VHDL Designs

Now we are ready to compile our design. Click the Smart button. This will compile all of the files that need to be recompiled. During the course of a design you might have many files and periodically make changes only to some subset of them. Rather than having to keep track of compilation order and the files that need to be recompiled, Galaxy keeps track of the modifications. The Selected button can be used recompile only selected files.

Once the compilation starts, a window will open with a stream of messages detailing progress of the compilation and any error messages. If there are errors in compilation, there is easy access to the editor from the Galaxy window. Successful compilation will be indicated in the status bar in the lower left portion of the window.

WARP does not have a VHDL simulator. However, it does have a functional simulator that will utilize the JEDEC file produced by the synthesis compiler and simulate the fitted design. In addition, our choice of synthesis options produced a VHDL model with accurate timing information. Check the project directory. You will find a subdirectory named \Vhd. In this directory you will find *structfa.vhd*. This is a VHDL simulation model compatible with the IEEE 1164 standard. The model has accurate timing information derived from knowledge of the hardware primitives of the device. In this case, we synthesized the design to a Cypress Semiconductor programmable logic device. The package provides the timing accurate simulation model. Look in this VHDL file and note the presence of the clause

library primitive;

use primitive.c37xp.all;

The preceding statements refer to packages that capture timing accurate models of the synthesized design. We can now use a VHDL simulator such as Speedwave, or the functional simulator in WARP called NOVA. The use of the former is described in Appendix A, whereas the latter is described in the following section.

Step 7: Invoke the Simulator

The WARP tools do not have a VHDL simulator, but rather provide a functional simulator that can simulate the JEDEC file produced by the synthesis process. This functional simulator is based on the JEDEC file that is used to program the programmable logic devices. Therefore the internal view of the simulator is based on the target device. The `Nova` simulator can be invoked from the `Galaxy` window from the `Tools` menu item.

Step 8: Load the Model

From the File menu item in the `Nova` window, select `Open`. The file *structfa.jed* should appear in the file selection dialog box. Select this file and click OK. The simulatable model has now been loaded. The Nova window will include node names on the left and the empty traces in the right part of the window.

Step 9: Creating View Nodes

Before we can begin simulating the design, we must familiarize ourselves with a few concepts and terminology used within `Nova`. Within a PLD we typically have regions of the circuit where we would like to trace signal values. This is captured in the idea of a *node* and *view node*. A node corresponds to the region of a circuit and this node can be configured so that we can trace a specific signal within the node region. For example, in a region corresponding to an output pin on the chip, we might actually trace the value of the signal before the output buffer or trace the value of the output enable. In WARP terminology we select a view of the region. The concepts of nodes and view nodes enable us to think about the signals we want to trace in terms of where they are in the target chip rather than their location in the logical architecture described by the VHDL model. If we are interested in simulation behavior, this is probably not the way we may wish to analyze the behavior of the design. If we are interested in implementation behavior, this would probably be the preferred manner in which we would like to study the behavior of the design.

From the `Views` menu item, select `Select View`. From the resulting dialog box select FULL and OK. The left side of the `Nova` window now will list all of the nodes in the design. Select one of the output signals from the chip, c_out by clicking on the button with the signal name on it. The trace should change color. Having selected this node, select `Create View Node` from the `Edit` menu item. The resulting dialog box is shown in Figure B-6. We can select one of several points on this node to trace, for example, the input to this node, which would be the output from the PLD array, or the output from this node. In general we need some knowledge of the physical layout within the chip to make informed choices when creating view nodes. This a problem with this view of the world if we are only interested in the simulation behavior. For our purposes, we are interested in the input and output signals of the model, and therefore select the node output.

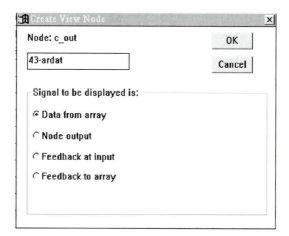

FIGURE B-6 Creating a view of a node

Step 10: Creating a View

When you simulate a circuit you may be interested in the behavior of a set of specific signals rather than looking at all of the signals in a circuit. This notion is captured through the definition of a *view*. A view is a set of signals that you wish to trace (or rather, in WARP's parlance, a set of nodes). You can create a view by selecting `Edit Views` from the `Views` menu item in the `Nova` window. The resulting dialog box will appear as shown in Figure B-7. On the left is the full set of view nodes in the circuit. On the right is the set of nodes that belong to this view of the model. You can create a new view, name the view, and make any set of nodes part of this view using the `ADD` and `Cut` buttons. Notice the number of nodes in the left half of the window. They have a format as *<number jed_node number>*. This naming convention is specific to this class of devices. The majority of these nodes are internal to the PLD. They may correspond to regions of the circuit around the input/output pins, or in the vicinity of the inputs and outputs of the switch array. When you delve into the details of the implementation of PLDs, such detailed knowledge would be desirable and very useful. For the moment, we are simply interested in fitting our design and ensuring that it is functionally correct. So we will ignore these internal nodes and concentrate on the inputs and outputs of the design entity that we have synthesized—the full adder.

Two default views are automatically created within the WARP environment. These are named `FULL`, and `PINS` and `REGS`. You can select the current view by selecting the `Select View` option from the `Views` menu item. Create a new view called `Entity` and include in this view the input and output signals of the full adder. Your resulting `Nova` trace window should appear as shown in Figure B-8.

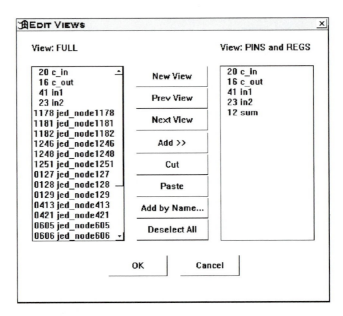

FIGURE B-7 The Edit Views dialog box

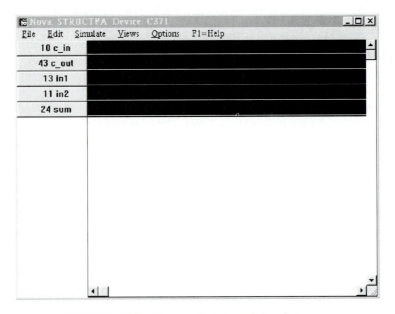

FIGURE B-8 The Nova window and the view

Step 11: Generating Input Stimulus

The next step is to provide a stimulus to the design and study the behavior. Select the node in1. From the Edit menu item select clock. The clock dialog box pops up. Accept the defaults except for the delay, which should be set to 10. When you accept this clock specification, a periodic waveform is generated on input in1. The clock signal is also delayed by 10 time units. Now, place the cursor at some point on the signal in2 and, while holding the left button down, drag the mouse some distance to the right. This portion of the trace should turn blue (or at least change color). From the Edit menu, select High. The portion of the trace of signal in2 that you selected should represent logic level 1. You should now see a pulse on input signal in2. Repeat this operation and generate pulses on other sections of the trace for in2 and also for the remaining input signal c_in. We have now generated an input stimulus on all of the input signals and are now ready to simulate the circuit.

Step 12: Running the Simulation

Select Execute from the Simulate Menu. The output waveforms will fill out, as shown in Figure B-9. Note the patterns that we generated on input signals in2 and c_in. From these patterns and the values of the output signals c_out and sum, we can verify

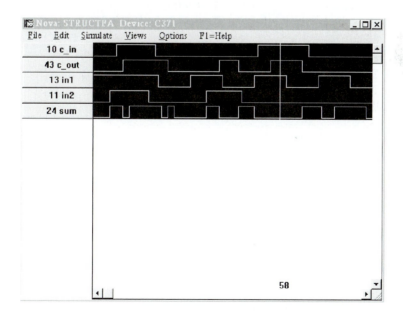

FIGURE B-9 A trace of a Nova simulation

that the full adder is functioning correctly. Using the horizontal scroll button on the bottom of the `Nova` window, pan to right of the trace. You will reach a region of the trace where there is only a periodic signal on input `in1`. The result will be a periodic signal on the sum `output`. Note the delay between the two signals. This represents the propagation delay from the input to the output for this implementation.

If you continue panning to the right you will notice that the trace stops at time 256. This is the default length of a *simulation segment*. You can change this length from the `Options` menu. Retain the `Nova` simulation for some miscellaneous exercises in the following steps.

Step 13: Time and Trace Resolution

You will have noticed in the `Nova` window that there is no logical time scale. You can create two *measuring cursors* for this purpose. Click on the bottom of the `Nova` window to generate the first cursor and click on the bottom while holding the shift–key down to generate the second cursor. The label at the bottom of the cursor indicates the time value at that point in the trace. You can move either measuring cursor by placing the mouse cursor over the time value, and moving the mouse with the left button depressed. You will notice the time values changing as you move the measuring cursor.

You can also change the resolution of the trace by selecting the `Resolution` option from the `Options` menu item. The resulting dialog box permits you to select the number of horizontal pixels devoted to a simulation clock tick. Set this number to 12. You will now notice that the trace has been "lengthened". The `Nova` window with the two measuring cursors and the lengthened trace will appear as shown in Figure B-10.

Step 14: Simulation Segments

You can create new segments by placing the first measuring cursor at some point and selecting `Create Segment` from the `Options` menu. This operation creates a new initial point for the simulation. Place the first measuring cursor at time 50 and create a new segment. Now select input `in1` and create a clock waveform on this signal with a period of 6, a high time of 3, and with no delay. You will notice that this waveform uses the beginning of the segment as the starting point, not time 0. Segments can deleted by selecting `Delete Segment` from the `Options` menu.

Step 15: Creating and Using Buses

It is sometimes useful to aggregate signals into a vector of signals which we will refer to as a bus. We can create such buses using the `Create Bus` option in the `Edit` menu. This will pop up a dialog box where you can select signals to be grouped into a bus and name

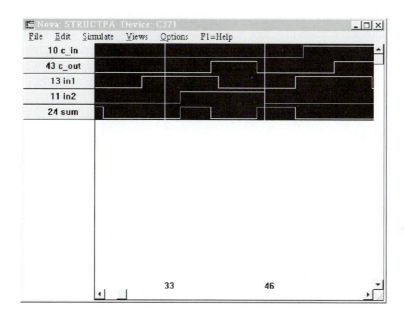

FIGURE B-10 Changing the resolution and using measuring cursors in the Nova window

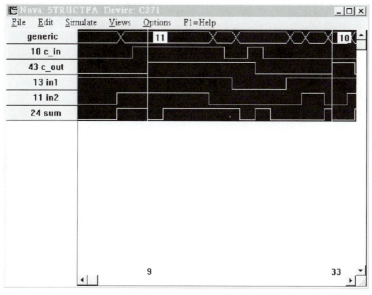

FIGURE B-11 Creation of Buses in Nova

the bus. Group the input signals in1 and in2 into a 2-bit signal named generic. The Nova window will now include this generic bus signal and will appear as shown in Figure B-11. By selecting from the Bus Radix option in the Edit menu item, you can select the number base with which to display the bus values. Select binary as the base for representing these values. The Nova display will appear as shown in Figure B-11.

Step 16: Exit

This tutorial has exercised only a subset of the features of the WARP tools. While covering the basic operations in the creation, synthesis, and simulation of VHDL models, the reader should be encouraged to explore the remaining capabilities of the toolset.

From the Nova window click on the File button and select Exit from the menu.

The SPIM Model: WVOffice

This Appendix contains the description of a structural model of the single-cycle SPIM datapath described in "Computer Organization & Design: The Hardware/Software Interface" by D. Patterson and J. Hennessy [9]. The model has been found useful as an application of VHDL to the modeling of computer architecture concepts as well as a tool for learning about computer architecture. Having to write or modify such models can reinforce architectural concepts from the classroom. This Appendix provides a brief description of the structure and organization of the model.

C.1 Model Overview

The single-cycle SPIM processor implements a basic reduced instruction set computer (RISC) style instruction set architecture and datapath. The model provided here supports a small set of instructions, limited memory, and a small set of registers. However, these characteristics can be easily modified. As shown in Figure C-1, the SPIM datapath is comprised of five components: fetch, decode, execute, memory, and control. Instruction memory is modeled within the fetch unit. The model implements five instructions, whose formats are shown in Figure C-2. The datapath is controlled by nine signals. The values of the control signals for realizing each instruction are computed in *control.vhd* (see Appendix C.8.6). Now that we know how to encode (supported) instructions, the next section describes the steps necessary to load and run this simulation model.

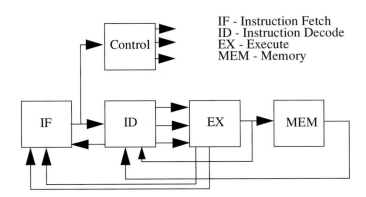

FIGURE C-1 Single-cycle SPIM VHDL model

Load /Store/Branch instructions

opcode	r1	r2	offset
6	5	5	16

Register Instructions

opcode	r1	r2	r3	shamt = 0	func
6	5	5	5	5	6

Jump Instruction

opcode	physical address
6	28

Instruction	Opcode	Operation
add r3, r1, r2	opcode = 0, func = 32	r3 = r1 + r2
sub r3, r1, r2	opcode = 0, func = 34	r3 = r1-r2
beq r1, r2, Loop	opcode = 4	if r1= r2, branch to Loop
j	opcode = 6	PC = address field
lw r2, (100)r1	opcode = 35	r2 = mem[r1+100]
sw r2, (100)r1	opcode = 43	memory[r1+100] = r2

FIGURE C-2 Instruction formats and operation of the simulation model

C.2 Sample Execution

To gain an understanding of how the model is organized and to ensure that you have a correct, functioning model, execute the following steps after creating all of the files in your project directory.

- Invoke the simulator you are using. Make sure you have the project set to your design directory.

- Ensure that the initialization files for the simulator you are using has the proper libraries. Note that the model uses the package `std_logic_1164`, which should be in library `IEEE`. The model uses the package `std_logic_arith`, which we have placed in library `WORK`, your working directory. **This is usually not the case!** You simply need to make sure that you reference this package in the correct location for the simulator that you are using. Refer to the simulator documentation or installation guide to find the correct location of this package.

- Invoke the VHDL analyzer and analyze **in order** each of the modules: *fetch.vhd, decode.vhd, execute.vhd, memory.vhd,* and *control.vhd.* These modules are provided in this appendix. After you have done this, analyze the top level structural module *ss_spim.vhd.*

- Note that the top level module has two input signals: `phi1` and `reset`. These signals must be provided stimulus from within the simulator. The datapath timing is set for an instruction execution time of 50 ns. Generate a clock signal as the stimulus for the `phi1` input with a period of 100 ns. Generate a single pulse of width 50 ns on the `reset` signal input.

- Open a trace window and select the set of signals that you wish to trace. Run the simulation for 800 ns and view the trace. This should produce the execution of the first 8 instructions stored in instruction memory in *fetch.vhd.* Verify that the model is behaving as expected.

- Examine the trace. This trace is an execution of a program such as the one shown below. Note that this program is first assembled, and then the words in instruction memory in *fetch.vhd* are initialized to this program.

```
L:   lw $2, 0($1)
     lw $4, 0($3)
     sub $4, $4, $2
     sw $4, 0($3)
     beq $4, $3, L
     and $4, $3, $4
     j L
     add $4, $5, $6
```

- Study the trace generated by the model given to you. The trace illustrates the instructions being executed, the outputs of the register file, the sign extended immediate operand, and some of the control signals. The cycle time of the datapath is 100 ns, that is, each instruction executes in 100 ns. Notice the relative delays in the datapath.

- You should be able to determine how to add other signals to trace or change the inputs to the datapath by examining your simulator documentation.

- By examining the files *fetch.vhd* and *memory.vhd* you will notice that the model supports only a small instruction memory and registers (i.e., only 8 words of instruction memory, 8 registers, and 8 words of data memory). Instruction memory is initialized to the program shown above. Data memory contents are initialized such that address 3 contains the value 0x33333333, address 4 contains 0x44444444, and so on. Registers are similarly initialized. You should now be able to interpret the trace. You can change any of these initial values as described below.

C.3 Sample Execution in Workview Office

To gain an understanding of how the model is organized and to ensure that you have a correct, functioning model, execute the following steps after copying all of the files into your project directory. These steps are for execution with Workview Office version 7.11.

- Invoke WVOffice. Make sure that the Project Manager has the project set to your design directory.

- Invoke Speedwave. Make sure that the VHDL Manager within Speedwave has the appropriate libraries selected (you should know about this from the tutorial in Appendix A). Note that the model uses the package std_logic_1164 which should be in library IEEE. The model also uses the package std_logic_arith which we have placed in library WORK, your working directory. **This is usually not the case!** You simply need to make sure that you reference this package in the correct location Refer to your local installation guide to find the correct location of this package.

- Invoke the VHDL analyzer and analyze **in order** each of the module: *fetch.vhd, decode.vhd, execute.vhd, memory.vhd,* and *control.vhd.* After you have done this, analyze the module *ss_spim.vhd.*

- Now you are ready to load and simulate the compiled model. Under the File menu, select the option to load the design. In the resulting dialog box, double click on your library unit. Select the entity *ss_spim.* Alternatively, perform the same operation from the VHDL Manager (under Tools). Click on your design library, select *ss_spim,* and then click on Simulate Obj.

- The model is now loaded, and the Navigator Window is also open within Speedwave. You can specify a command file now by selecting the option from the File menu item. Command files are sources of Speedwave commands that you frequently employ. Rather than run through the complete sequence each time you simulate the model, commands can be placed in a command file. The command file that is provided in Appendix C.8.7 will set up and run a sample simulation for 8 clock cycles and display a trace of program execution. Resize your windows to suit your viewing needs. Remember, the set of signals that are traced and the duration of the trace can be modified, by you or from within the simulator.

- Note that the top level module has two input signals: `phi1` and `reset`. These signals must be provided stimulus from within the simulator. The datapath timing is set for an instruction execution time of 50 ns. The command files generate a clock signal as the stimulus for the `phi1` input with a period of 100 ns, and a single pulse of width 50 ns on the `reset` signal input.

- Make the Trace window larger and examine the trace. This trace is an execution of a program such as the one shown below. Note that this program is first assembled, and then the words in instruction memory in *fetch.vhd* are initialized to this program.

L: lw $2, 0($1)

lw $4, 0($3)

sub $4, $4, $2

sw $4, 0($3)

beq $4, $3, L

and $4, $3, $4

j L

add $4, $5, $6

- Study the trace generated by the model given to you. The trace illustrates the instructions being executed, the outputs of the register file, the sign extended immediate operand, and some of the control signals. The step time is 50 ns and the cycle time of the datapath is also 100 ns. Thus, each instruction executes in 100 ns. Notice the relative delays in the datapath.

- Examine the command file to understand the steps taken. You should be able to determine how to add other signals to trace or change the inputs to the datapath by examining the command file.

- By examining the files *fetch.vhd*, and *memory.vhd*, you will notice that the model supports only a small instruction memory and registers: only 8 words of instruction memory, 8 registers, and 8 words of data memory. Instruction memory is initialized to the program shown above. Data memory contents are initialized such that address 3 contains the value 0x33333333, address 4 contains 0x44444444, and so forth. Registers are similarly initialized. You should now be able to interpret the trace. You can change any of these initial values as described below.

- Note that not all signals are traced. Change the command file *ss_spim.cmd* or provide commands at the `Speedwave` window to trace additional signals or remove old ones.

C.4 Loading a New Program into Memory

Note that the instruction memory is modeled in the instruction fetch unit which is described in *fetch.vhd*. Therefore, if you wanted to change the program being executed by this datapath, you would follow these steps:

1. Hand assemble your program. Remember that it should contain less than eight instructions unless you increase the size of memory.

2. Edit the *fetch.vhd* program to initialize the contents of the memory array to your assembled program.

3. Reanalyze *fetch.vhd* and *ss_spim.vhd*.

C.5 Extending the Datapath

Add new instructions to the datapath by following these steps:

1. Determine the operations to be performed to execute the instruction.

2. Decide which module(s) will implement the desired functionality.

3. Add any new signals that may be necessary.

4. Modify the computations in each module. **Test each module separately!** This can be done by simply clocking the inputs and checking the values of the outputs using the trace windows.

5. Make any necessary changes to *ss_spim.vhd*.

6. Load a new test program into *fetch.vhd*.

7. Reanalyze all *affected* modules, with *ss_spim.vhd* being analyzed last. Follow the rules presented in Section 8.2.

8. Load and execute the model as described above.

C.6 Some Helpful Pointers

- If signals appear as undefined, that is, XXXXXXXX, this may be because they are not correctly connected. Check *ss_spim.vhd*.

- If memory addresses are out of range—greater than eight—the address is computed modulo 8. For example if a fetch is initiated from address 10 (remember there are only 8 locations), then the address used is 10 *modulo* 8 = 2.

C.7 Datapath

Figure C-3 illustrates the detailed schematic of the Single-Cycle Datapath. The names of the signals are as declared in the model and can be found in *ss_spim.vhd*. These signals are used to interconnect the five basic components described in VHDL (*fetch.vhd, decode.vhd, execute.vhd, memory.vhd* and *control.vhd*). The specification of the interconnection of these components using the named signals can be found in *ss_spim.vhd*. For example, the IF module fetches the instruction for the ID module. This interconnection is realized by connecting the output of IF and the input of ID to the signal s_instruction. By examining the rest of *ss_spim.vhd* and following the schematic shown in Figure C-3, you should be able to determine how the schematic in the figure is created by statements in *ss_spim.vhd*. This will enable you to add any new signals that you may need to add. Note that in the figure control signals are shaded, while signals carrying data are darker. Furthermore, while the clock signal is distributed to all modules, it is used only in *fetch.vhd* to determine when the next instruction is to be fetched. In another implementation (e.g., multicycle) you would use the phi1 in each module. In this single cycle implementation, once an instruction is fetched, all of the other modules are driven to compute new outputs as a function of its inputs. The delays for the individual modules are set such that an instruction execution completes in one clock cycle. To provide an example of the structural code, *ss_spim.vhd* is included in Section C.8.1. This file describes the interconnection of the behavioral models as shown in Figure C-3.

C.8 VHDL Source for the SPIM Model

The first section provides the top-level structural model that describes the interconnection between all of the datapath components. The following sections contain the behavioral models of each component. The last section provides a command file for Viewlogic's Workview Office 7.11 simulator. This command file sets up the display and runs the simulation for 8 clock cycles. This SPIM model uses the library std_logic_arith compiled into the library WORK. You might otherwise expect to find this package in the library IEEE. The files that use this package should be edited to note the correct source on the system that you are using.

C.8.1 ss_spim.vhd

```
-- Top level SPIM module
--
-- As for signal naming conventions, all the signals used to connect the
-- modules are prefixed with "s_". For example s_MemToReg is the signal used
-- to connect the MemToReg control output from the control unit to the blocks
-- that require this control signal.
--
```

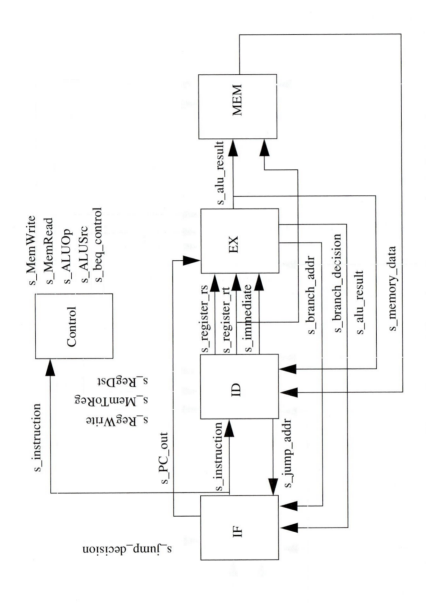

FIGURE C-3 Structural model of the single-cycle SPIM datapath

-- branch addresses and the branch decision are computed in the execute module
-- and provided to the fetch module

```
Library IEEE;
use IEEE.std_logic_1164.all;
use WORK.std_logic_arith.all;

entity ss_spim is
port(phi1, reset: in std_logic);
end ss_spim;

architecture structural of ss_spim is
--
-- Instruction fetch unit
--
component fetch
generic(module_delay: Time := 10 ns);
port(PC_out, instruction : out std_logic_vector(31 downto 0);
    branch_addr, jump_addr: in std_logic_vector(31 downto 0);
    branch_decision,jump_decision, reset, phi1 : in std_logic);
end component;
--
-- Instruction decode
--
component decode
generic(module_delay: Time:= 10 ns);
port(
    instruction : in std_logic_vector(31 downto 0);                    -- inputs
    memory_data, alu_result :in std_logic_vector(31 downto 0);
    RegDst, RegWrite, MemToReg, reset, phi1 : in std_logic;
--
    register_rs, register_rt, jump_addr :out std_logic_vector(31 downto 0);    -- outputs
    immediate :out std_logic_vector(31 downto 0));
end component;
-- Execution unit
--
component execute
generic (module_delay: Time:= 10 ns);
port(
    PC4, register_rs, register_rt :in std_logic_vector(31 downto 0);    -- inputs
    immediate :in std_logic_vector(31 downto 0);
    ALUOp: in std_logic_vector(1 downto 0);
    ALUSrc, beq_control, phi1 : in std_logic;

    branch_addr, alu_result :out std_logic_vector(31 downto 0);    -- outputs
    branch_decision : out std_logic);
end component;
--
```

```
-- Data Memory
--
component memory
generic (module_delay : Time := 10 ns);
port(
    address, write_data : in std_logic_vector(31 downto 0); -- inputs
    MemWrite, MemRead : in std_logic;
    phi1 : in std_logic;

    read_data :out std_logic_vector(31 downto 0));        --- outputs
end component;
--
-- Control
--
component control
generic(module_delay :Time := 10 ns);
port(instruction : in std_logic_vector(31 downto 0);
    reset : in std_logic;
    RegDst, MemRead, MemToReg, MemWrite :out std_logic;
    ALUSrc, RegWrite, Jump, beq_control: out std_logic;
    ALUOp: out std_logic_vector(1 downto 0));
end component;
--
-- global 32 bit signals
--
signal s_branch_addr, s_PC_out:std_logic_vector(31 downto 0);
signal s_immediate : std_logic_vector(31 downto 0);
signal s_instruction, s_alu_result : std_logic_vector(31 downto 0);
signal s_register_rs, s_register_rt : std_logic_vector(31 downto 0);
signal s_memory_data, s_jump_addr: std_logic_vector(31 downto 0);

--   global single bit signals

signal s_RegDst, s_MemRead, s_MemWrite, s_MemToReg,s_ALUSrc:std_logic;
signal s_RegWrite, s_branch_decision, s_jump_decision, s_beq_control :std_logic;
signal s_ALUOp :std_logic_vector(1 downto 0);
--
-- configuration
--
for IFE: fetch use entity WORK.fetch(behavioral);
for ID : decode use entity WORK.decode(behavioral);
for EX: execute use entity WORK.execute(behavioral);
for MEM: memory use entity WORK.memory(behavioral);
for spim_control: control use entity WORK.control(behavioral);

begin
IFE: fetch
port map(PC_out => s_PC_out,
```

```
            instruction => s_instruction,
            branch_addr => s_branch_addr,
            jump_addr => s_jump_addr,
            jump_decision => s_jump_decision,
            branch_decision => s_branch_decision,
            reset =>reset,
            phi1 => phi1);
      ID: decode
      port map(instruction => s_instruction,
            memory_data =>s_memory_data,
            alu_result => s_alu_result,
            RegDst => s_RegDst,
            RegWrite => s_RegWrite,
            MemToReg => s_MemToReg,
            reset => reset,
            phi1 => phi1,
            register_rs => s_register_rs,
            register_rt => s_register_rt,
            immediate => s_immediate,
            jump_addr => s_jump_addr);
      EX: execute
      port map(PC4 => s_PC_out,
            register_rs => s_register_rs,
            register_rt => s_register_rt,
            immediate => s_immediate,
            ALUOp => s_ALUOp,
            ALUSrc => s_ALUSrc,
            phi1 =>phi1,
            branch_addr => s_branch_addr,
            alu_result => s_alu_result,
            branch_decision => s_branch_decision,
            beq_control => s_beq_control);
      MEM: memory
      port map(address => s_alu_result,
            write_data => s_register_rt,
            MemWrite =>s_MemWrite,
            MemRead => s_MemRead,
            phi1 => phi1,
            read_data => s_memory_data);
      spim_control: control
      port map(instruction => s_instruction,
            reset => reset,
            RegDst => s_RegDst,
            MemRead => s_MemRead,
            MemToReg => s_MemToReg,
            MemWrite => s_MemWrite,
            ALUSrc => s_ALUSrc,
            RegWrite => s_RegWrite,
```

```
        ALUOp => s_ALUOp,
        Jump => s_jump_decision,
        beq_control => s_beq_control);
end structural;
```

C.8.2 fetch.vhd

```
-- Instruction fetch behavioral model. Instruction memory is provided within this model.
-- IF increments the PC, and writes the appropriate output signals after module_delay.
-- The next value of the PC (i.e., branch or PC+1) is determined in the EX stage.
-- Therefore, IF simply uses the input value of PC to access instruction memory.
-- This value may be PC+1 depending upon the branch condition

Library IEEE;
use IEEE.std_logic_1164.all;
use WORK.std_logic_arith.all;

entity fetch is
generic(module_delay: Time := 10 ns);
port(PC_out, instruction : out std_logic_vector(31 downto 0);
    branch_addr, jump_addr: in std_logic_vector(31 downto 0);
    branch_decision,jump_decision, reset, phi1 : in std_logic);
end fetch;

architecture behavioral of fetch is
type mem_array is array(0 to 7) of std_logic_vector(31 downto 0);
                            -- you can make memory larger by
                            -- increasing the size of the array
begin
process
-------------------------------------------------------------------------
-- Instruction memory is an array of 32 bit words that are initialized with
-- the hand assembled instruction values

-------------------------------------------------------------------------

variable   mem : mem_array := (
        to_stdlogicvector(X"8c220000"),    --- L: lw $2, 0($1)
        to_stdlogicvector(X"8c640000"),    ---    lw $4, 0($3)
        to_stdlogicvector(X"00822022"),    ---    sub $4, $4, $2
        to_stdlogicvector(X"ac640000"),    ---    sw $4, 0($3)
        to_stdlogicvector(X"1022fffb"),    ---    beq $1, $2, L
        to_stdlogicvector(X"00642824"),    ---    and $4, $3, $4
        to_stdlogicvector(X"18000001"),    ---    j L
        to_stdlogicvector(X"00a62020"));   ---    add $4, $5, $6

variable PC : std_logic_vector(31 downto 0);
variable temp_result : std_logic_vector(31 downto 0);
```

```
variable index : integer := 0;

begin
 --  wait until the start of a cycle
 wait until (phi1'event and phi1 = '1');
 if reset = '1' then                        -- check if reset, if so initialize the PC
      PC :=to_stdlogicvector(X"00000000");    -- & is the concatenation operator
                                             -- this produces a 33 bit number
      instruction <= to_stdlogicvector( X"00000000");
      wait until reset = '0' and phi1 = '1';    --  wait until next clock
 end if;
 -- check if we are to take the branch address, jump address or just the incremented value of
 -- of the PC
 if (branch_decision = '1') then
      PC := branch_addr;
 end if;
 if (jump_decision = '1') then
      PC := jump_addr;
 end if;
 --
 -- determine address of instruction. Since we only have 8 words in instruction
 -- memory we only use (need!) the three least significant bits of the address
 -- we use 4 to account for two's complement representations
 index := to_integer(PC(3 downto 0));

 -- increment PC for access of next instruction. Note here we only add 1 rather
 -- than 4 since we have implemented instruction memory as an array of
 -- words. Remember that variable values are persistent, i.e., variables retain
 -- their values the next time this process is executed.

 PC := PC + to_stdlogicvector(x"1");
 --send instruction to decode and execute stages

 PC_Out <= PC after module_delay;
 instruction <= mem(index) after module_delay;
 end process;
end behavioral;
```

C.8.3 decode.vhd

```
--
-- instruction decode unit.
--
-- this model consists of two concurrent processes - 1) one for initialization on reset and register file
-- write operation and 2) one for performing register file reads. The register file is structure that is
-- shared by both processes. Note that this model only provides 8 registers!
--
```

```
Library IEEE;
use IEEE.std_logic_1164.all;
use WORK.std_logic_arith.all;

entity decode is
generic(module_delay: Time:= 10 ns);
port(
    instruction : in std_logic_vector(31 downto 0);
    memory_data, alu_result :in std_logic_vector(31 downto 0);
    RegDst, RegWrite, MemToReg, reset, phi1 : in std_logic;
    register_rs, register_rt, jump_addr:out std_logic_vector(31 downto 0);
    immediate :out std_logic_vector(31 downto 0));
end decode;

architecture behavioral of decode is
signal reg0, reg1, reg2, reg3, reg4 : std_logic_vector(31 downto 0);
signal reg5, reg6, reg7 : std_logic_vector(31 downto 0);
begin
---------------------------------------------------------------------------
-- Process to write the register file when necessary
---------------------------------------------------------------------------
reg_write: process(reset,memory_data,alu_result)
variable addr1,addr2, addr3: std_logic_vector(4 downto 0);
variable write_value :std_logic_vector(31 downto 0);
begin
-- on reset initialize the register values
    if reset = '1' then
    reg0  <= to_stdlogicvector(X"00000000");
    reg1  <= to_stdlogicvector(X"11111111");
    reg2  <= to_stdlogicvector(X"22222222");
    reg3  <= to_stdlogicvector(X"33333333");
    reg4  <= to_stdlogicvector(X"44444444");
    reg5  <= to_stdlogicvector(X"55555555");
    reg6  <= to_stdlogicvector(X"66666666");
    reg7  <= to_stdlogicvector(X"77777777");
  else
--   determine address of register to be written, i.e., rt or rd?
 addr2 := instruction(20 downto 16);
 addr3 := instruction(15 downto 11);
 if RegDst = '0' then
     addr3 := addr2;
 end if;
 --
 -- Determine source operand to be written, i.e., memory or alu result
 --
 if RegWrite = '1' then
     if MemToReg = '1' then
         write_value := memory_data;
```

```
      else
            write_value := alu_result;
      end if;

      case addr3 is
      when "00000" =>
            reg0 <= to_stdlogicvector(x"00000000");
      when "00001" =>
            reg1 <= write_value;
      when "00010" =>
            reg2 <= write_value;
      when "00011" =>
            reg3 <= write_value;
      when "00100" =>
            reg4 <= write_value;
      when "00101" =>
            reg5 <= write_value;
      when "00110" =>
            reg6 <= write_value;
      when "00111" =>
            reg7 <= write_value;
      when others =>
            reg0 <= to_stdlogicvector(x"00000000");
      end case;
   end if;
end if;
end process reg_write;
---------------------------------------------------------------------------
-- Process to read the register file and pass the operands to the execute
-- unit
---------------------------------------------------------------------------
reg_access: process(instruction)
variable rt, rs, imm : std_logic_vector(31 downto 0);
variable addr1,addr2 : std_logic_vector(4 downto 0);
variable neg :std_logic_vector(15 downto 0) := to_stdlogicvector(x"FFFF");
variable pos :std_logic_vector(15 downto 0) := to_stdlogicvector(x"0000");

begin
                  -- compute register addresses for indexing the
                  -- register file array
addr1 := instruction(25 downto 21);
addr2 := instruction(20 downto 16);
                  -- access registers
      case addr1 is
      when "00000" =>
            rs := reg0;
      when "00001" =>
            rs := reg1;
```

```
        when "00010" =>
             rs := reg2;
        when "00011" =>
             rs := reg3;
        when "00100" =>
             rs := reg4;
        when "00101" =>
             rs := reg5;
        when "00110" =>
             rs := reg6;
        when "00111" =>
             rs := reg7;
        when others =>
             rs:= to_stdlogicvector(X"ffffffff");     -- if register read returns this value you
                                                       -- know you have a register out of range
    end case;

        case addr2 is
        when "00000" =>
             rt := reg0;
        when "00001" =>
             rt := reg1;
        when "00010" =>
             rt := reg2;
        when "00011" =>
             rt := reg3;
        when "00100" =>
             rt := reg4;
        when "00101" =>
             rt := reg5;
        when "00110" =>
             rt := reg6;
        when "00111" =>
             rt := reg7;
        when others =>
             rt:= to_stdlogicvector(X"ffffffff");     -- if register read returns this value you
                                                       -- know you have a register out of range
    end case;
                    -- access immediate from the instruction
imm(15 downto 0) := instruction(15 downto 0);

                    -- perform sign extension
if instruction(15) = '1' then
     imm(31 downto 16) := neg;
else
     imm(31 downto 16) := pos;
end if;
-- determine if the instruction is a jump instruction if so compute the jump address.
```

```
-- The IF unit will worry about whether to use it or not. The following line computes
-- the shifted jump address. Remember, since the simulation uses word addressed
-- memory, jump addresses are not multiplied by 4!
--
jump_addr(31 downto 0) <= to_stdlogicvector(B"000000") & instruction(25 downto 0);
--
-- set output signals
--
register_rs <= rs after module_delay;
register_rt <= rt after module_delay;
immediate <= imm after module_delay;

end process reg_access;
end behavioral;
```

C.8.4 execute.vhd

```
--
-- execution unit. Only a subset of instructions are supported in this
-- model, specifically add, sub, lw, sw, beq, and, or
--
Library IEEE;
use IEEE.std_logic_1164.all;
use WORK.std_logic_arith.all;

entity execute is
generic (module_delay: Time:= 10 ns);
port(
    PC4, register_rs, register_rt :in std_logic_vector(31 downto 0);
    immediate :in std_logic_vector(31 downto 0);
    ALUOp: in std_logic_vector(1 downto 0);
    ALUSrc, beq_control, phi1 : in std_logic;
    branch_addr, alu_result :out std_logic_vector(31 downto 0);
    branch_decision : out std_logic);
end execute;

architecture behavioral of execute is

begin
process(register_rs, register_rt, immediate)
variable branch_offset : std_logic_vector(31 downto 0) := to_stdlogicvector(X"00000000");
variable temp_branch_addr : std_logic_vector(31 downto 0);
variable alu_output : std_logic_vector(31 downto 0);
            -- this variable is 33 bits long since it contains the
            -- results of addition operations which can produce
            -- 33 bit results
variable zero : std_logic := '0'; -- control bit that indicates if ALU output = 0
```

```
begin
-- assuming the instruction is a branch, compute the branch address
-- shift by 2 bits left. This value will be required if the instruction is
-- indeed a branch
branch_offset := immediate; -- recall we do not multiply by 4 in this model
                -- since we model a word addressed memory rather
                -- than a byte addressed memory

temp_branch_addr := PC4 + branch_offset;
--
-- Depending upon the specific code (ALUOp)
-- perform the corresponding alu function
--
case ALUOp is
    when "00" =>
--
-- this is a memory instruction so simply compute the memory address
-- and set the zero bit accordingly
--
    alu_output := register_rs + immediate;

    if (alu_output = to_stdlogicvector(x"00000000")) then   -- set Z bit is result is 0
        zero := '1';
    else
        zero := '0';
    end if;
    when "01" =>
--
-- this is a branch instruction
--
    alu_output := register_rs - register_rt;
    if (alu_output = to_stdlogicvector(x"00000000")) then   -- set Z bit is result is 0
        zero := '1';
    else
        zero := '0';
    end if;
    when "10" =>
--
-- these are alu operations that need the function code. Recall this is the
-- 6 least significant bits of the instruction which are available here as the
-- 6 least significant bits of the immediate operand.
-- based on these bits, perform ADD, SUB, AND, OR
--

    case immediate(5 downto 0) is
    when "100000" =>                -- add
        alu_output := register_rs + register_rt;
```

```
        when "100010" =>              --  subtract
            alu_output := register_rs - register_rt;

        when "100100" =>              --  AND instruction
            alu_output := register_rs and register_rt;

        when "100101" =>              --  OR instruction
            alu_output := register_rs or register_rt;

        when others =>
--
-- in this case set error codes
--
        alu_output := to_stdlogicvector(X"ffffffff");
        end case;

        if (alu_output = to_stdlogicvector(x"00000000")) then    -- set Z bit is result is 0
            zero := '1';
        else
            zero := '0';
        end if;

        when others =>
        alu_output := to_stdlogicvector(X"ffffffff");
end case;
--
-- output signal assignments
--
branch_decision <= (beq_control and zero) after module_delay;
branch_addr <= temp_branch_addr after module_delay;
alu_result <= alu_output after module_delay;

end process;
end behavioral;
```

C.8.5 memory.vhd

```
--
-- data memory component.
--
Library IEEE;
use IEEE.std_logic_1164.all;
use WORK.std_logic_arith.all;

entity memory is
generic (module_delay : Time := 10 ns);
```

```vhdl
port(
    address, write_data : in std_logic_vector(31 downto 0);
    MemWrite, MemRead : in std_logic;
    phi1 : in std_logic;
    read_data :out std_logic_vector(31 downto 0));
end memory;

architecture behavioral of memory is
type mem_array is array(0 to 7) of std_logic_vector(31 downto 0);

begin
process(address,write_data)
variable data_mem : mem_array := (
    to_stdlogicvector(X"00000000"),    --- initialize data memory
    to_stdlogicvector(X"11111111"),    --- to some values
    to_stdlogicvector(X"22222222"),    ---
    to_stdlogicvector(X"33333333"),
    to_stdlogicvector(X"44444444"),
    to_stdlogicvector(X"55555555"),
    to_stdlogicvector(X"66666666"),
    to_stdlogicvector(X"77777777"));
variable addr :integer;
variable mem_contents : std_logic_vector(31 downto 0);

begin
addr := to_integer(address(2 downto 0));    --  since we have only a memory with
                    --  8 words, we use only the three
                    --  least significant bits of the address
mem_contents := write_data;

if MemWrite = '1' then
    data_mem(addr) := mem_contents;
elsif MemRead = '1' then
    mem_contents := data_mem(addr);
    read_data <= mem_contents after module_delay;
end if;
end process;
end behavioral;
```

C.8.6 control.vhd

```vhdl
--
-- control unit. simply implements the truth table for a small set of
-- instructions
--
Library IEEE;
use IEEE.std_logic_1164.all;
```

```vhdl
use WORK.std_logic_arith.all;

entity control is
generic(module_delay :Time := 5 ns);
port(instruction : in std_logic_vector(31 downto 0);
    reset : in std_logic;
    RegDst, MemRead, MemToReg, MemWrite :out std_logic;
    ALUSrc, RegWrite, Jump, beq_control: out std_logic;
    ALUOp: out std_logic_vector(1 downto 0));
end control;

architecture behavioral of control is

begin
process(instruction)
variable rformat, lw, sw, beq, jmp, addiu :std_logic; -- define local variables
                    -- corresponding to instruction
                    -- type
variable opcode : std_logic_vector(5 downto 0);

begin
if reset = '1' then
    RegDst <= '0';
    ALUSrc <= '0';
    MemToReg <= '0';
    RegWrite <= '0';
    MemRead <= '0';
    MemWrite <= '0';
    ALUOp(1 downto 0) <=  to_stdlogicvector(B"00");
    Jump <= '0';
    beq_control <= '0';
else
--
-- recognize opcode for each instruction type
--
 opcode(5 downto 0) := instruction(31 downto 26);

rformat := ((NOT opcode(5)) AND (NOT opcode(4)) AND (NOT
opcode(3)) AND (NOT opcode(2)) AND (NOT opcode(1))
AND (NOT opcode(0)));

lw := (( opcode(5)) AND (NOT opcode(4)) AND (NOT
opcode(3)) AND (NOT opcode(2)) AND opcode(1) AND
opcode(0));

sw := (( opcode(5)) AND (NOT opcode(4)) AND
opcode(3) AND (NOT opcode(2)) AND opcode(1) AND
opcode(0));
```

beq := ((NOT opcode(5)) AND (NOT opcode(4)) AND
(NOT opcode(3)) AND opcode(2) AND (NOT opcode(1))
AND (NOT opcode(0)));

jmp := ((NOT opcode(5)) AND (NOT opcode(4)) AND
(NOT opcode(3)) AND opcode(2) AND opcode(1)
AND (NOT opcode(0)));

addiu := ((NOT opcode(5)) AND (NOT opcode(4)) AND
opcode(3) AND (NOT opcode(2)) AND (NOT opcode(1))
AND opcode(0));
--
-- basically implement each output signal as the column of the truth
-- table on pg. 296 which defines the control
--

RegDst <= rformat after module_delay;
ALUSrc <= (lw or sw or addiu) after module_delay;
MemToReg <= lw after module_delay;
RegWrite <= (rformat or lw or addiu) after module_delay;
MemRead <= lw after module_delay;
MemWrite <= sw after module_delay;
ALUOp(1 downto 0) <= (rformat & beq) after module_delay;
Jump <= jmp after module_delay;
beq_control <= beq after module_delay;
end if;
end process;
end behavioral;

C.8.7 Workview Office Command File

The following command file for Workview office can be invoked once the model is loaded
into the simulator. The command file will generate stimulus for the clock and reset inputs,
open a trace window, and run the simulation for 800 ns. This file can be modified to suit
your own simulation and viewing needs. Note how comments are delineated.

```
|-----------------------------------------------------------------------------
|-- command file for Workview Office 7.11
|-----------------------------------------------------------------------------

stepsize 50ns
wfm /reset @0ns=1 @50ns =0
clock /phi1 1 0
|--
|-- these commands change the radix of 32 bits numbers to hexadecimal notation
|-- to make it easier to display on the trace
```

```
|--
radix hex /s_instruction
radix hex /s_register_rs
radix hex /s_register_rt
radix hex /s_alu_result
radix hex /s_immediate
radix hex /s_memory_data
radix hex /s_pc_out
|--
|-- open a trace window and trace the following signals
|--
wave ss_spim.wfm /phi1 /reset /s_instruction /s_pc_out /s_regwrite /s_register_rs /s_register_rt /
s_aluop
wave ss_spim.wfm /s_alu_result /s_memory_data /s_regdst
|--
|-- run the simulation for 800 ns
|--
sim 800ns
|--
|-- end command file
|--
```

Standard VHDL Packages

This appendix contains listings of interfaces to standard packages available with the VHDL distributions. The STANDARD and TEXTIO packages are provided as part of the implementation of the VHDL environment. The package std_logic_1164 is an implementation of the IEEE 1164 value system and is generally provided by most, if not all, vendors to support the generation of portable VHDL models. There are several other packages that are also typically made available by vendors. The reader is encouraged to browse through the package headers at your installation and study the contents.

D.1 Package STANDARD

The package STANDARD is distributed by all vendors. This package provides the definitions of the predefined types and functions for the language. The package header contents shown here is from the 1993 IEEE Standard VHDL Language Reference Manual [5]. The implementation of this package will be consistent across all vendors. The listing of the package header shown below omits the definition of all of the operators for each type.

```
-------------------------------------------------------------------------------
-- ANSI/IEEE Std 1076–1993
-- IEEE Standard VHDL Language Reference Manual
-- Copyright ©1993 by the Institute of Electrical and Electronics Engineers, Inc.
-- The IEEE disclaims any responsibility or liability resulting from the placement
-- and use in the described manner. Information is reprinted with the permission
-- of the IEEE
--
-------------------------------------------------------------------------------
--
--Predefined enumeration types
--
package STANDARD is
   type BOOLEAN is (FALSE, TRUE);
   type BIT is ('0', '1');
   type CHARACTER is (
      NUL, SOH, STX, ETX, EOT, ENQ, ACK, BEL,
      BS, HT, LF, VT, FF, CR, SO, SI,
      DLE, DC1, DC2, DC3, DC4, NAK, SYN, ETB,
      CAN, EM, SUB, ESC, FSP, GSP, RSP, USP,
      ' ', '!', '"', '#', '$', '%', '&', ''',
      '(', ')', '*', '+', ',', '-', '.', '/',
      '0', '1', '2', '3', '4', '5', '6', '7',
      '8', '9', ':', ';', '<', '=', '>', '?',
      '@', 'A', 'B', 'C', 'D', 'E', 'F', 'G',
      'H', 'I', 'J', 'K', 'L', 'M', 'N', 'O',
      'P', 'Q', 'R', 'S', 'T', 'U', 'V', 'W',
      'X', 'Y', 'Z', '[', '\', ']', '^', '_',
      '`', 'a', 'b', 'c', 'd', 'e', 'f', 'g',
      'h', 'i', 'j', 'k', 'l', 'm', 'n', 'o',
      'p', 'q', 'r', 's', 't', 'u', 'v', 'w',
      'x', 'y', 'z', '{', '|', '}', '~', DEL);
--
-- there are a host of other characters here including some special characters which
-- are omitted from this presentation
--
   type SEVERITY_LEVEL is (NOTE, WARNING, ERROR, FAILURE);
   type universal_integer  is range implementation_defined;
   type universal_real  is range implementation_defined;
--
-- in implementations of this package the statement below would define
-- numeric values in the range field
--
   type INTEGER is range implementation_defined;
```

type REAL **is range** *implementation_defined*;
type TIME **is range** *implementation_defined*
 units
   fs;   -- femtosecond
   ps = 1000 fs; -- picosecond
   ns = 1000 ps; -- nanosecond
   us = 1000 ns; -- microsecond
   ms = 1000 us; -- millisecond
   sec = 1000 ms; -- second
   min = 60 sec; -- minute
   hr = 60 min; -- hour
 end units;
subtype DELAY_LENGTH **is** TIME **range** 0 fs **to** TIME'HIGH;
-- function that returns the current simulation time:
impure function NOW **return** DELAY_LENGTH;
 -- predefined numeric subtypes:
 subtype NATURAL **is** INTEGER **range** 0 **to** INTEGER'HIGH;
 subtype POSITIVE **is** INTEGER **range** 1 **to** INTEGER'HIGH;
 -- predefined array types:
 type STRING **is array** (POSITIVE **range** <>) **of** CHARACTER;
 type BIT_VECTOR **is array** (NATURAL **range** <>) **of** BIT;
--
--predefined types for opening files
--
type FILE_OPEN_KIND **is** (READ_MODE, WRITE_MODE, APPEND_MODE);
type FILE_OPEN_STATUS **is** (OPEN_OK, STATUS_ERROR, NAME_ERROR, MODE_ERROR);

attribute FOREIGN: STRING;

end STANDARD;

D.2 Package TEXTIO

The TEXTIO package is distributed by all vendors. This package provides the definitions of the predefined types and functions of the language for performing input/output operations on text files. The package header contents shown here is from the 1993 IEEE Standard VHDL Language Reference Manua [5]l. The implementation of this package will be consistent across all vendors. The listing of the package header shown below omits the definition of the operators for each type.

```
---------------------------------------------------------------------------
-- ANSI/IEEE Std 1076-1993
-- IEEE Standard VHDL Language Reference Manual
-- Copyright ©1993 by the Institute of Electrical and Electronics Engineers, Inc.
-- The IEEE disclaims any responsibility or liability resulting from the placement
-- and use in the described manner. Information is reprinted with the permission
-- of the IEEE
--
---------------------------------------------------------------------------
```

package TEXTIO **is**
-- Type Definitions for Text I/O
--

    type LINE **is access** STRING;   -- A LINE is a pointer to a STRING value.
    type TEXT **is file of** STRING;   -- A file of variable-length ASCII records.
    type SIDE **is** (RIGHT, LEFT);   -- For justifying output data within fields.
    subtype WIDTH **is** NATURAL;   -- For specifying widths of output fields.

--
-- Standard text files:
-- Note these are different from VHDL'87
--

    file INPUT: TEXT **open** READ_MODE **is** "STD_INPUT";
    file OUTPUT: TEXT **open** WRITE_MODE **is** "STD_OUTPUT";

--
-- Input routines for standard types:

    procedure READLINE (**file** F: TEXT; L: **out** LINE);
    procedure READ (L: **inout** LINE; VALUE: **out** BIT;   GOOD: **out** BOOLEAN);
    procedure READ (L: **inout** LINE; VALUE: **out** BIT);
    procedure READ (L: **inout** LINE; VALUE: **out** BIT_VECTOR; GOOD: **out** BOOL-
EAN);
    procedure READ (L: **inout** LINE; VALUE: **out** BIT_VECTOR);
    procedure READ (L: **inout** LINE; VALUE: **out** CHARACTER; GOOD: **out** BOOLEAN);
    procedure READ (L: **inout** LINE; VALUE: **out** CHARACTER);
    procedure READ (L: **inout** LINE; VALUE: **out** INTEGER; GOOD: **out** BOOLEAN);
    procedure READ (L: **inout** LINE; VALUE: **out** INTEGER);
    procedure READ (L: **inout** LINE; VALUE: **out** REAL; GOOD: **out** BOOLEAN);
    procedure READ (L: **inout** LINE; VALUE: **out** REAL);
    procedure READ (L: **inout** LINE; VALUE: **out** STRING; GOOD: **out** BOOLEAN);
    procedure READ (L: **inout** LINE; VALUE: **out** STRING);
    procedure READ (L: **inout** LINE; VALUE: **out** TIME; GOOD: **out** BOOLEAN);
    procedure READ (L: **inout** LINE; VALUE: **out** TIME);

-- Output routines for standard types

```
    procedure WRITELINE  (file F: TEXT; L: inout LINE);
    procedure WRITE  (L: inout LINE; VALUE: in BIT;
            JUSTIFIED: in SIDE := RIGHT; FIELD: in WIDTH := 0);
    procedure WRITE  (L: inout LINE; VALUE: in BIT_VECTOR;
            JUSTIFIED: in SIDE := RIGHT; FIELD: in WIDTH := 0);
    procedure WRITE  (L: inout LINE; VALUE: in BOOLEAN;
            JUSTIFIED: in SIDE := RIGHT; FIELD: in WIDTH := 0);
    procedure WRITE  (L: inout LINE; VALUE: in CHARACTER;
            JUSTIFIED: in SIDE := RIGHT; FIELD: in WIDTH := 0);
    procedure WRITE  (L: inout LINE; VALUE: in INTEGER;
            JUSTIFIED: in SIDE := RIGHT; FIELD: in WIDTH := 0);
    procedure WRITE  (L: inout LINE; VALUE: in REAL;
 JUSTIFIED: in SIDE := RIGHT; FIELD: in WIDTH := 0; DIGITS: in NATURAL:=0);
    procedure WRITE  (L: inout LINE; VALUE: in STRING;
            JUSTIFIED: in SIDE := RIGHT; FIELD: in WIDTH := 0);
    procedure WRITE  (L: inout LINE; VALUE: in TIME;
        JUSTIFIED: in SIDE := RIGHT; FIELD: in WIDTH := 0; UNIT : in TIME;= ns);
end TEXTIO;
```

D.3 The Standard Logic Package

This package defines the types and supporting functions for the implementation of the IEEE 1164 value system. It is made available by most if not all vendors and is placed in the library IEEE.

```
-- ------------------------------------------------------------------------------------------------
-- IEEE Std 1164–1993
-- IEEE Standard Multivalue Logic System for VHDL Model Interoperability
-- Copyright © 1993 by the Institute of Electrical and Electronics Engineers, Inc.
-- The IEEE disclaims any responsibility or liability resulting from the placement
-- and use in the described manner. Information is reprinted with the permission
-- of the IEEE
--
-- ------------------------------------------------------------------------------------------------
--    Title        : std_logic_1164 multi-value logic system
--    Library      : This package shall be compiled into a library
--                 : symbolically named IEEE.
--                 :
--    Developers   : IEEE model standards group (par 1164)
--    Purpose      : This packages defines a standard for designers
--                 : to use in describing the interconnection data types
--                 : used in vhdl modeling.
--                 :
```

```
--  Limitation      : The logic system defined in this package may
--                  : be insufficient for modeling switched transistors,
--                  : since such a requirement is out of the scope of this
--                  : effort. Furthermore, mathematics, primitives,
--                  : timing standards, etc. are considered orthogonal
--                  : issues as it relates to this package and are therefore
--                  : beyond the scope of this effort.
--                  :
--  Note            : No declarations or definitions shall be included in,
--                  : or excluded from this package. The "package declaration"
--                  : defines the types, subtypes and declarations of
--                  : std_logic_1164. The std_logic_1164 package body shall be
--                  : considered the formal definition of the semantics of
--                  : this package. Tool developers may choose to implement
--                  : the package body in the most efficient manner available
--                  : to them.
--                  :
-- -----------------------------------------------------------------
--  modification history :
-- -----------------------------------------------------------------
-- version | mod. date:|
--  v4.200 | 01/02/91 |
-- -----------------------------------------------------------------
```

PACKAGE `std_logic_1164` **IS**

```
----------------------------------------------------------------
-- logic state system (unresolved)
----------------------------------------------------------------
```

TYPE std_ulogic **IS** ('U', -- Uninitialized
                    'X', -- Forcing Unknown
                    '0', -- Forcing 0
                    '1', -- Forcing 1
                    'Z', -- High Impedance
                    'W', -- Weak   Unknown
                    'L', -- Weak   0
                    'H', -- Weak   1
                    '-' -- Don't care
                );

```
----------------------------------------------------------------
-- unconstrained array of std_ulogic for use with the resolution function
----------------------------------------------------------------
```

TYPE std_ulogic_vector **IS ARRAY** (NATURAL **RANGE** <>) **OF** std_ulogic;

```
----------------------------------------------------------------
-- resolution function
```

FUNCTION resolved (s : std_ulogic_vector) **RETURN** std_ulogic;

-- *** industry standard logic type ***

SUBTYPE std_logic **IS** resolved std_ulogic;

-- unconstrained array of std_logic for use in declaring signal arrays

----

TYPE std_logic_vector **IS ARRAY** (NATURAL **RANGE** <>) **OF** std_logic;

-- common subtypes

SUBTYPE X01   **IS** resolved std_ulogic **RANGE** 'X' **TO** '1'; -- ('X','0','1')
SUBTYPE X01Z   **IS** resolved std_ulogic **RANGE** 'X' **TO** 'Z'; -- ('X','0','1','Z')
SUBTYPE UX01   **IS** resolved std_ulogic **RANGE** 'U' **TO** '1'; -- ('U','X','0','1')
SUBTYPE UX01Z **IS** resolved std_ulogic **RANGE** 'U' **TO** 'Z'; --
('U','X','0','1','Z')

-- overloaded logical operators

FUNCTION "and" (l : std_ulogic; r : std_ulogic) **RETURN** UX01;
FUNCTION "nand" (l : std_ulogic; r : std_ulogic) **RETURN** UX01;
FUNCTION "or"   (l : std_ulogic; r : std_ulogic) **RETURN** UX01;
FUNCTION "nor" (l : std_ulogic; r : std_ulogic) **RETURN** UX01;
FUNCTION "xor" (l : std_ulogic; r : std_ulogic) **RETURN** UX01;
-- function "xnor" (l : std_ulogic; r : std_ulogic) return ux01;
FUNCTION "not" (l : std_ulogic   ) **RETURN** UX01;

-- vectorized overloaded logical operators

FUNCTION "and" (l, r : std_logic_vector) **RETURN** std_logic_vector;
FUNCTION "and" (l, r : std_ulogic_vector) **RETURN** std_ulogic_vector;
FUNCTION "nand" (l, r : std_logic_vector) **RETURN** std_logic_vector;
FUNCTION "nand" (l, r : std_ulogic_vector) **RETURN** std_ulogic_vector;
FUNCTION "or"   (l, r : std_logic_vector) **RETURN** std_logic_vector;
FUNCTION "or"   (l, r : std_ulogic_vector) **RETURN** std_ulogic_vector;
FUNCTION "nor" (l, r : std_logic_vector) **RETURN** std_logic_vector;
FUNCTION "nor" (l, r : std_ulogic_vector) **RETURN** std_ulogic_vector;
FUNCTION "xor" (l, r : std_logic_vector) **RETURN** std_logic_vector;
FUNCTION "xor" (l, r : std_ulogic_vector) **RETURN** std_ulogic_vector;

-- ---

-- Note : The declaration and implementation of the "xnor" function is
-- specifically commented until at which time the VHDL language has been

-- officially adopted as containing such a function. At such a point,
-- the following comments may be removed along with this notice without
-- further "official" ballotting of this std_logic_1164 package. It is
-- the intent of this effort to provide such a function once it becomes
-- available in the VHDL standard.
-- --
-- function "xnor" (l, r : std_logic_vector) return std_logic_vector;
-- function "xnor" (l, r : std_ulogic_vector) return std_ulogic_vector;
   FUNCTION "not" (l : std_logic_vector) **RETURN** std_logic_vector;
   FUNCTION "not" (l : std_ulogic_vector) **RETURN** std_ulogic_vector;

   -- conversion functions

   FUNCTION To_bit   (s : std_ulogic;   xmap : BIT := '0') **RETURN** BIT;
   FUNCTION To_bitvector (s : std_logic_vector ; xmap : BIT := '0') **RETURN**
BIT_VECTOR;
   FUNCTION To_bitvector (s : std_ulogic_vector; xmap : BIT := '0') **RETURN**
BIT_VECTOR;
   FUNCTION To_StdULogic   (b : BIT   ) **RETURN** std_ulogic;
   FUNCTION To_StdLogicVector (b : BIT_VECTOR) **RETURN** std_logic_vector;
   FUNCTION To_StdLogicVector (s : std_ulogic_vector) **RETURN** std_logic_vector;
   FUNCTION To_StdULogicVector (b : BIT_VECTOR   ) **RETURN**
std_ulogic_vector;
   FUNCTION To_StdULogicVector (s : std_logic_vector) **RETURN**
std_ulogic_vector;

   -- strength strippers and type convertors

   FUNCTION To_X01 (s : std_logic_vector) **RETURN** std_logic_vector;
   FUNCTION To_X01 (s : std_ulogic_vector) **RETURN** std_ulogic_vector;
   FUNCTION To_X01 (s : std_ulogic   ) **RETURN** X01;
   FUNCTION To_X01 (b : BIT_VECTOR   ) **RETURN** std_logic_vector;
   FUNCTION To_X01 (b : BIT_VECTOR   ) **RETURN** std_ulogic_vector;
   FUNCTION To_X01 (b : BIT   ) **RETURN** X01;
   FUNCTION To_X01Z (s : std_logic_vector) **RETURN** std_logic_vector;
   FUNCTION To_X01Z (s : std_ulogic_vector) **RETURN** std_ulogic_vector;
   FUNCTION To_X01Z (s : std_ulogic   ) **RETURN** X01Z;
   FUNCTION To_X01Z (b : BIT_VECTOR   ) **RETURN** std_logic_vector;
   FUNCTION To_X01Z (b : BIT_VECTOR   ) **RETURN** std_ulogic_vector;
   FUNCTION To_X01Z (b : BIT   ) **RETURN** X01Z;
   FUNCTION To_UX01 (s : std_logic_vector) **RETURN** std_logic_vector;
   FUNCTION To_UX01 (s : std_ulogic_vector) **RETURN** std_ulogic_vector;
   FUNCTION To_UX01 (s : std_ulogic   ) **RETURN** UX01;
   FUNCTION To_UX01 (b : BIT_VECTOR   ) **RETURN** std_logic_vector;

FUNCTION To_UX01 (b : BIT_VECTOR   ) **RETURN** std_ulogic_vector;
FUNCTION To_UX01 (b : BIT   ) **RETURN** UX01;

-- edge detection

FUNCTION rising_edge (**SIGNAL** s : std_ulogic) **RETURN** BOOLEAN;
FUNCTION falling_edge (**SIGNAL** s : std_ulogic) **RETURN** BOOLEAN;

-- object contains an unknown

FUNCTION Is_X (s : std_ulogic_vector) **RETURN** BOOLEAN;
FUNCTION Is_X (s : std_logic_vector) **RETURN** BOOLEAN;
FUNCTION Is_X (s : std_ulogic   ) **RETURN** BOOLEAN;
END std_logic_1164;

D.4 Other Useful Packages

In addition, users should be aware that vendors will provide other packages that encapsulate many useful functions such as those for arithmetic, for handling real numbers, as well as miscellaneous utility functions such as type conversion. Some of these packages are vendor specific, while others are currently subject to efforts to arrive at some standards of use.

Check the vendor documentation for other packages that may be available on your system. It is useful to browse through the package headers to obtain an idea of the sets of functions, procedures, or data types that are supported within these packages and thereby understand the motivation for their development. Packages for mathematical functions and type conversion are perhaps the first themes that come to mind. When we think in terms of hardware design, several other needs also become evident. For example, given a specific gate-level design, we may have packages containing various implementation alternatives for the same set of components. One package may contain models for the high-speed implementation of the components, while another may contain models corresponding to the low-power implementation of the same components. The structuring mechanism provided by packages is put to good use in creating libraries of component models used within an organization. Often these packages are proprietary products of the parent organization.

A Starting Program Template

This appendix serves as a quick reference guide to the structure of a first VHDL model. A template for a general VHDL model is presented. This template can help with the syntactical arrangement of programming constructs. It is useful when trying to remember where to place statements within a program relative to other program constructs. This template can serve as a handy reference for quickly constructing our first VHDL models. The goal here is to provide a template that contains the most basic and common (and therefore, for our purposes, important) language features and will enable the reader to rapidly proceed to the construction of useful VHDL models. This template follows VHDL'87.

We can combine the procedures for constructing behavioral and structural models that are described in the early chapters and identify a sequence of common operations. The first step is the construction of a schematic of the system being modeled.

Construct_Schematic

1. Represent each component (e.g., gate) of the system to be modeled as a *delay element.* The delay element simply captures all of the delays associated with the computation represented by the component and propagation of signals through the component. For each output signal of a component associate a specific value of delay through the component for that output signal.

2. Draw a schematic interconnecting all of the components. Uniquely label each component.

3. Identify the input signals of the system as input ports.

4. Identify the output signals of the system as output ports.

5. All remaining signals are internal signals and should be uniquely labeled.

6. Associate a type, such as **bit**, **bit_vector**, or `std_logic_vector`, with each input port, output port, and internal signal.

7. Ensure that each input port, output port, and internal signal is labeled with a unique name.

This schematic can now be translated into a VHDL model containing behavioral and structural models of the components that comprise the system. In fact, the architecture body shown in Figure E-1 can be structured as a series of program statements. Each statement can be one of the following:

1. A concurrent signal assignment statement

- simple signal assignment

- conditional signal assignment

- selected signal assignment

2. A process

The process may have a sensitivity list, and may comprise a large block of sequential code. Recall that a process execution takes no simulation time and may produce events on signals that are scheduled at some time in the future.

3. A component instantiation statement

If components have been declared in addition to signals, these components may be instantiated and their input and output ports mapped to signals declared in the architecture. In this manner, these components can be "connected" to, or communicate with, other components, CSAs, or processes.

This leads to the following procedure for constructing general models reflecting the behavior of the digital system.

Construct_Behavioral_Model

1. At this point I recommend using the IEEE 1164 value system. To do so, include the following two lines at the top of your model declaration.

        library IEEE;
        use IEEE.std_logic_1164.all;

Single-bit signals can be declared to be of type `std_logic` while multibit quantities can be declared to be of type `std_logic_vector`.

2. Select a name for the entity (`entity_name`) for the system and write the entity description specifying each input or output signal port, its mode, and associated type.

3. Select a name for the architecture (`arch_name`) and write the architecture description. Within the architecture description, name and declare all of the internal signals used to connect the components. These signal names are shown on your schematic. The architecture declaration states the type of each signal and possibly an initial value.

4. For each delay element decide if the behavior of the block will be described by concurrent signal assignment statements, processes, or a component instantiation statement. Depending upon the type, perform the following:

   4.1 *CSA*: For each output signal of the component select a concurrent signal assignment statement that expresses the value of this signal as a function of the signals that are inputs to that component. Use the value of the propagation delay through the component provided for that output signal. The output signal and/or one or more input signals may be a port of the entity.

   4.2 *Process*: Alternatively, if the computation of the signal values at the outputs of the component are too complex to represent with concurrent signal assignment statements, describe the behavior of the component with a process. One or more processes can be used to compute the values of the output signals from that component. For each process perform the following:

   4.2.1 Label the process. If you are using a sensitivity list, identify the signals that will activate the process.

   4.2.2 Declare variables used within the process.

   4.2.3 Write the body of the process computing the values of output signals and the relative time at which these output signals assume these values. If a sensitivity list is not used, specify wait statements at appropriate points in the process to specify when the process should suspend and when it should resume execution. It is an error to have both a sensitivity list and a wait statement within the process.

   4.2.4 Complete the process with a set of signal assignment statements, assigning the computed values to the output signals. These output signals may be signals internal to the architecture or may be port signals found in the entity description.

   4.3 *Component Instantiation*: For those components for which entity–architecture pairs exist.

   4.3.1 Construct component declarations for each unique component that will be used in the model. A component declaration can be easily constructed from the component's entity description. For example, the port list is identical.

   4.3.2 Within the declarative region of the architecture description (i.e., before the **begin** statement), list the component declarations.

   4.3.3 Within the declarative region of the architecture description (i.e., before the **begin** statement), list the configuration specification if not using the default binding for the component entities.

```
library IEEE;
use IEEE.std_logic_1164.all;
```
— Declare libraries and packages

```
entity entity_name is
port (input signals : in type;
      output signals : out type);
end entity_name;
```
— Interface description

```
architecture arch_name of entity_name is
component C1
generic (constant : type :=-- initialization;
port (input signals : in type;
      output signals : out type);
end component;

component C2
generic (constant : type :=-- initialization;
port (input signals : in type;
      output signals : out type);
end component;
```
— Component declaration

— Architecture declarative region

```
-- declare signals used to interconnect
-- behavioral and structural models

signal s1, s2, s3 : type:= initialization;
```
— Signal declaration

```
-- configuration specifications
for L1: C1   use entity --identify entity–architecture;
for L2: C2   use entity -- identify entity–architecture;
```
— Configuration specification

```
begin
  -- CSA, process, or component instantiation statement
  -- CSA, process, or component instantiation statement
  -- CSA, process, or component instantiation statement
  -- CSA, process, or component instantiation statement
end arch_name;
```
— Architecture body

FIGURE E-1 Anatomy of a VHDL model

4.3.4 Write the component instantiation statement. The label is derived from the schematic followed by the **port map** construct. The port map statement will have as many entries as there are ports on the component. If necessary, include a **generic map** statement.

5. If there are signals that are driven by more than one source, the type of this signal must be a resolved type. This type must have a resolution function declared for use with signals of this type. For our purposes use the IEEE 1164 types `std_logic` for single-bit signals and `std_logic_vector` for bytes, words, or multibit quantities. These are resolved types. Make sure you include the **library** clause and the **use** clause to include all of the definitions provided in the `std_logic_1164` package.

6. If you are using any functions or type definitions provided by a third party make sure that you have declared the appropriate library using the **library** clause and declared the use of this package via the presence of a **use** clause in your model.

These steps will produce a fairly generic model. In particular, this approach implies that all design units (entity, architecture, and configuration information) is placed in one physical file. This is clearly not necessary. For example, we know from Chapter 5 that configurations are distinct design units that may be described separately. However, it is often easier to start in the fashion shown here. As our expertise grows, we will be able to avail ourselves of the advantages of dealing with design units separately and managing them effectively. Finally, note that in Figure E-1 the location of the packages is shown as library `IEEE`. Depending on the packages, this may not be the case, and when writing models we must have knowledge of the location of any vendor-supplied packages that are being used. We may also be creating our own libraries for retaining user packages.

Index